Topics in Neuroscience

M. Filippi • M. Rovaris • G. Comi (Eds)

Neurodegeneration in Multiple Sclerosis

Springer

Massimo Filippi
Neuroimaging Research Unit
Department of Neurology
Scientific Institute and
University Ospedale San Raffaele
Milan, Italy

Giancarlo Comi
Department of Neurology
Scientific Institute and
University Ospedale San Raffaele
Milan, Italy

Marco Rovaris
Neuroimaging Research Unit
Department of Neurology
Scientific Institute and
University Ospedale San Raffaele
Milan, Italy

The Editors and Authors wish to thank Bayer Healthcare–Bayer Schering Pharma for the support and help in the realization and promotion of this volume

ISBN 978-88-470-5583-4 Springer Milan Berlin Heidelberg New York
ISBN 978-88-470-0391-0 (eBook)

Springer-Verlag is a part of Springer Science+Business Media
springer.com

Softcover re-print of the Hardcover 1st edition 2007

Typesetting: C & G di Cerri e Galassi, Cremona, Italy
Printing and binding: Grafiche Porpora, Segrate (MI), Italy
Cover design: Simona Colombo, Milan, Italy

Springer-Verlag Italia S.r.l.,Via Decembrio 28, I-20137 Milan

Introduction

Multiple sclerosis (MS) leads to the formation of macroscopic, discrete foci of tissue damage in the central nervous system (CNS). These lesions can be seen on conventional magnetic resonance imaging (MRI) scans, making this technique a sensitive tool for diagnosing MS and for monitoring its evolution, although it is incapable of disentangling the heterogeneous features of MS lesion pathology, which may range from edema to permanent loss of myelin and axons. Moreover, it is now well known that MS does not spare the normal-appearing white (NAWM) and gray (NAGM) matter, i.e., those portions of the CNS which appear intact on conventional MRI scans. Normal-appearing tissue changes seem to be either secondary to intrinsic damage caused by T2-visible lesions (via Wallerian degeneration of fibers passing through macroscopic abnormalities) or the result of independent pathological processes. In the NAWM, the main pathological findings are gliosis, microglial activation, disturbances of the blood-brain barrier, and also demyelination and loss of axons. In the NAGM, less inflammatory changes are seen, but numerous lesions can be identified ex vivo, which may be associated with irreversible axonal and neuronal loss. All these findings indicate that MS does not have to be considered a purely inflammatory condition, but rather a disease where inflammation and neurodegeneration play complementary pathogenetic roles.

The past 10 years have seen continuous advances in MS treatment. Following the approval of interferon β-1b as a disease-modifying therapy for relapsing-remitting MS, other immunomodulating and immunosuppressive treatments have demonstrated significant success in reducing the activity of the disease, in terms of clinical relapses and MRI lesions. Nevertheless, this treatment efficacy is not accompanied by an equal ability to prevent or slow down the progressive clinical deterioration which occurs in MS, even independently of acute relapses. It is conceivable that the accumulation of MS-related neurological disability is secondary to the neurodegenerative components of MS pathology and, as a consequence, ad hoc therapeutic strategies are warranted. For this reason, future MS trials will need reliable in vivo surrogates of neurodegeneration in order to better assess the efficacy of treatments aimed at preventing its evolution. In this context, several MR-based techniques have been investigated as tools capable of providing reliable pieces of information on neurodegeneration in MS.

MR-based measurements of CNS "atrophy" represent a useful tool to assess the final outcome of neurodegeneration. The poor short- to medium-term correlation between atrophy and T2-visible lesion load, which has been consistently reported by several studies, supports the notion that tissue volume reductions may primarily reflect "occult" neurodegeneration and give complementary information to that provided by conventional MRI. The assessment of the burden of T1 "black holes" is another, easily implementable, technique to quantify the extent of MRI-visible tissue disruption, which is related to a decrease in axonal loss. Measuring the deposition of iron, as reflected by T2 relaxation time abnormalities, can also provide estimates of the extent of MS neurodegeneration in clinically eloquent brain areas, such as the basal ganglia and the cortical GM. Among more sophisticated MR-based methodologies, both magnetization transfer (MT) and diffusion tensor (DT) MRI are now widely applied in the study of MS. Correlative studies have confirmed that a significant relationship exists between decreased MT ratio or increased diffusivity and increased loss of myelin and axons, both within and outside focal MS lesions; thereby making these techniques potential candidates for monitoring trials of neurodegeneration in MS. Proton magnetic resonance spectroscopy (^{1}H-MRS) has the unique advantage that it can provide information with a high biochemical specificity for ongoing tissue changes. Among ^{1}H-MRS-derived metabolic measures, the levels of N-acetylaspartate (NAA) represent a highly specific correlate of neuronal and axonal viability. The information provided by structural MR-based techniques on the neurodegenerative components of MS pathology can be integrated with those coming from functional MRI (fMRI) studies. With the latter technique, the ability of the MS brain to limit the consequences of irreversible tissue damage can be explored. fMRI data indicate that cortical reorganization in MS patients begins soon after the clinical onset of the disease and continues through the entire course of the disease.

Although new MRI modalities are likely to provide us with more specific in vivo measures reflecting the neurodegenerative features of MS pathology, their application to clinical trial monitoring requires a careful preliminary consideration of several methodological issues, including the need to achieve a satisfactory trade-off between pathological specificity, sensitivity to longitudinal changes, and the feasibility of their use in the setting of large-scale, multicenter studies. In other words, MR-derived measures of neurodegeneration, although promising, need to be properly validated as surrogates of MS. In this context, useful lessons can be learned by studies of other neurodegenerative conditions, as well as by the application of other paraclinical biomarkers.

With this book, we aim to provide a complete and up-to-date overview of technical, methodological, and clinical issues related to the application of MRI in MS trials of neurodegeneration. This review is the result of an international workshop held in Milan on 10 June 2005, during the Ninth Annual Advanced Course on the Use of Magnetic Resonance Techniques in Multiple Sclerosis, and subsequent discussions among the authors. We hope that, for clinicians and researchers, this book will be of help for setting up the scenario they will have to face when assessing the efficacy of future therapeutic strategies in the treatment of MS.

Milan, March 2007

M. Filippi
M. Rovaris
G. Comi

Table of Contents

Background

MRI Techniques to Assess Neurodegeneration

Evaluation of MRI Outcomes

Lessons from Other Neurodegenerative Diseases

Designing MS Trials for Neurodegeneration

List of Contributors

Inmaculada Aban
Department of Biostatistics
University of Alabama at Birmingham
Birmingham, AL, USA

Ashish Arora
Brigham & Women's Hospital
Harvard Medical School
Boston, MA, USA

Rohit Bakshi
Brigham & Women's Hospital
Harvard Medical School
Boston, MA, USA

Elisabetta Capello
Department of Neuroscience
Ophtalmology and Genetics
University of Genoa
Genoa, Italy

Declan T. Chard
NMR Research Unit
Department of Neuroinflammation
Institute of Neurology
University College London
London, UK

Giancarlo Comi
Department of Neurology
Scientific Institute and
University Ospedale San Raffaele
Milan, Italy

Gary Cutter
Department of Biostatistics
University of Alabama at Birmingham
Birmingham, AL, USA

Venkata S.R. Dandamudi
Brigham & Women's Hospital
Harvard Medical School
Boston, MA, USA

Nicola De Stefano
Department of Neurological
and Behavioral Sciences
University of Siena
Siena, Italy

Massimo Filippi
Neuroimaging Research Unit
Department of Neurology
Scientific Istitute and
University Ospedale San Raffaele
Milan, Italy

Giovanni B. Frisoni
LENITEM - Laboratory of Epidemiology
Neuroimaging & Telemedicine
IRCCS San Giovanni di Dio
The National Center for Research
and Care of Alzheimer's Disease
Brescia, Italy

Yulin Ge
Department of Radiology
Center for Biomedical Imaging
New York University School of Medicine
New York, NY, USA

Gavin Giovannoni
Department of Neuroscience
Institute of Cell and Molecular Science
Queen Mary University of London
London, UK

Sharmilee Gnanapavan
Department of Neuroimmunology
Institute of Neurology
London, UK

Robert I. Grossman
Department of Radiology
Center for Biomedical Imaging
New York University School of Medicine
New York, NY, USA

Matilde Inglese
Department of Radiology
Center for Biomedical Imaging
New York University School of Medicine
New York, NY, USA

Meng Law
Department of Radiology
Center for Biomedical Imaging
New York University School of Medicine
New York, NY, USA

Letizia Leocani
Department of Neurology
Scientific Institute and
University Ospedale San Raffaele
Milan, Italy

Giovanni Luigi Mancardi
Department of Neuroscience
Ophtalmology and Genetics
University of Genoa
Genoa, Italy

Mario Mascalchi
Radiodiagnostic Section
Department of Clinical Physiopathology
and Department of Neurological Sciences
University of Florence
Florence, Italy

David H. Miller
NMR Research Unit
Department of Neuroinflammation
Institute of Neurology
University College London
London, UK

Kim Mouridsen
Centre for Functionally Integrative
Neuroscience (CFIN)
Department of Neuroradiology
Århus University Hospital
Århus C, Denmark

Mohit Neema
Brigham & Women's Hospital
Harvard Medical School
Boston, MA, USA

Leif Østergaard
Centre for Functionally Integrative
Neuroscience (CFIN)
Department of Neuroradiology
Århus University Hospital
Århus C, Denmark

Elisabetta Perego
Neuroimaging Research Unit
Scientific Institute and
University Ospedale San Raffaele
Milan, Italy

Matteo Pizzorno
Department of Neuroscience
Ophtalmology and Genetics
University of Genoa
Genoa, Italy

Waqar Rashid
NMR Research Unit
Department of Neuroinflammation
Institute of Neurology
University College London
London, UK

Maria A. Rocca
Neuroimaging Research Unit
Department of Neurology
Scientific Istitute and
University Ospedale San Raffaele
Milan, Italy

Marco Rovaris
Neuroimaging Research Unit
Department of Neurology
Scientific Istitute and
University Ospedale San Raffaele
Milan, Italy

Per Soelberg Sørensen
Danish Multiple Sclerosis Research Center
Department of Neurology
Copenhaguen University Hospital
Copenhaguen, Denmark

Maria Pia Sormani
Biostatistics Unit
Department of Health Sciences (DISSAL)
University of Genova
Genoa, Italy

James Stankiewicz
Brigham & Women's Hospital
Harvard Medical School
Boston, MA, USA

Antonio Uccelli
Department of Neuroscience
Ophtalmology and Genetics
University of Genoa
Genoa, Italy

BACKGROUND

Chapter 1

Neuropathological Advances in Multiple Sclerosis

E. Capello, A. Uccelli, M. Pizzorno, G.L. Mancardi

Introduction

The neuropathology of multiple sclerosis (MS) has been thoroughly and carefully described since the pioneer studies on the disease [1]. It was already clear in the last years of the 1800s that MS is an inflammatory disease of the central nervous system (CNS) with scattered areas of demyelination and relative sparing of axons. The astroglial and microglial reactions, the possible loss of neurons, the axonal injury, and the capacity of the CNS to partly remyelinate the damaged areas were all widely known and described.

The model of experimental allergic encephalomyelitis (EAE), first described in primates [2] and then fully developed in rodents, suggested the possible pathogenesis of the disease, widely accepted by the entire neurological community. According to the EAE model, MS is an autoimmune acquired disorder of the white matter of the CNS. The antigen is a myelin protein or other proteins of the CNS; depending on the genetic background of the patient and for some still unidentified environmental and external factors, some T lymphocytes become sensitized, cross the blood-brain barrier, reach the target antigen presented by the local microglia, and damage the nervous tissue. The first hit is followed by the appearance in the CNS of other inflammatory cells, not necessarily activated against antigens of the CNS, which contribute to and amplify the damage. The inflammatory infiltrate, made up by CD4 and CD8 lymphocytes and by macrophages, is responsible for the myelin loss in a specific area of the CNS. Inflammation is then followed by the astroglial reaction, partial repair of the injured tissue, which is often extensive [3], and a decrease in the inflammatory activity. For some unknown reasons the disease can become chronic and other waves from the blood to the CNS follow, with the appearance of other areas of demyelination and inflammation in different parts of the CNS. With time, the CNS becomes colonized by inflammatory cells which remain inside the brain and spinal cord, contributing to the progressive damage of the tissue and the progression of the clinical symptoms.

This classical view of the pathogenesis of the disease has been partly revised in recent years, due to the discovery of new neuropathological data, which indicate that MS is a considerably more complicated disorder, in which multiple mechanisms of damage of the CNS are probably involved.

Axonal Damage

Axonal damage and the loss of axons and neurons were identified early, in the first neuropathological studies of the disease, but it is only recently that axonal changes (Fig. 1) have been emphasized and considered as the neuropathogical counterpart of the non-reversible neurological disability [4, 5]. Numerous studies have demonstrated that axonal damage, shown by the accumulation of amyloid precursor protein (APP) in axons when axonal transport is impaired, occurs early in the course of the disease [6] or in the acute plaques in the EAE model [7]. However, with time, axons are progressively lost, and in old sclerotic plaques more than 60% of the axons have disappeared [8]. Axons can be damaged by cytokines, activated T cells, glutamate, nitric oxide, or other substances present in the inflammatory process. It has recently been demonstrated that CD8+ T cells are more represented than CD4+ lymphocytes in the demyelinated areas. The axonal damage is related especially to the presence of CD8+ T cells and clonal expansion is more prominent in this T cell population, while CD4+ T cells are mainly located in the perivascular space [9].

Damaged axons, devoid of myelin, cannot survive for very long, and progressively degenerate. The loss of axons not protected by myelin is a well-known neurobiologic phenomenon, which can also occur without inflammation, as it

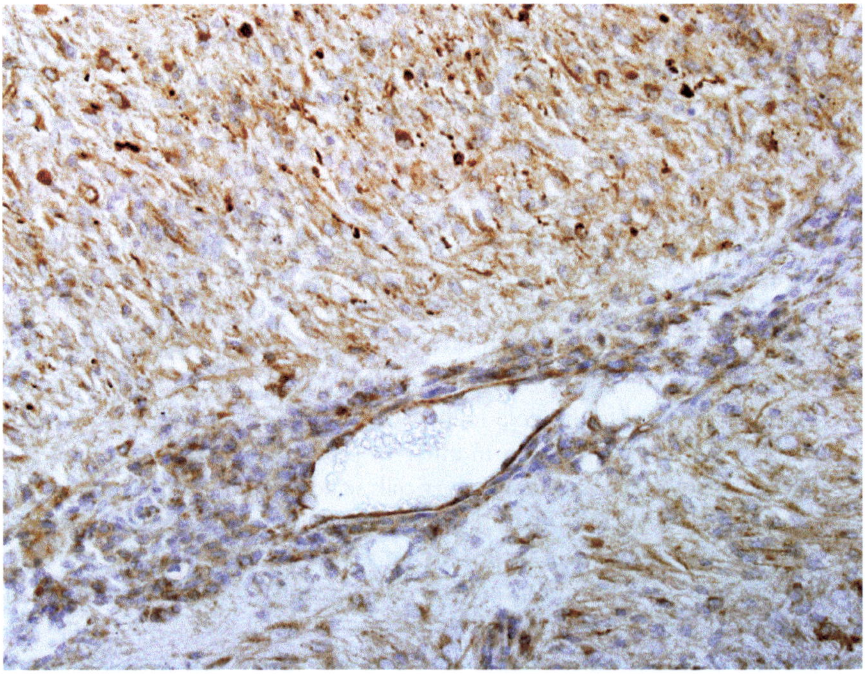

Fig. 1. SMI32-positive axons in active plaque (40×)

occurs in hereditary myelin disorders [10], or in "in vitro" models of demyelination [11], where axons without the envelopment of the myelin sheath show morphological and neurochemical changes indicative of stress, damage and degeneration. Accumulation of neuronal type voltage-gated calcium channels or a redistribution of ion channels along the demyelinated axon may contribute to the axonal dysfunction and subsequently to the axonal loss. Moreover, a defect in the Na channel aggregation may be involved in the remyelination failure [12]. One of the main targets for the immediate future is to discover the reasons for the progressive axonal degeneration after myelin has disappeared and then to develop neuroprotective drugs to prevent the process of progressive axonal loss.

The Role of B Lymphocytes

The attention of many researchers has been directed in recent years to the study and characterization of the B cells. This is not surprising, considering that the presence in the cerebrospinal fluid (CSF) of oligoclonal bands made by IgG is the most important laboratory marker of MS. B cells, plasmacells and IgG directed against myelin proteins are present in the demyelinated areas [13], and antimyelin antibodies, when detected in the CSF, may have prognostic value [14]. The "pattern II" neuropathology described by Lucchinetti et al. [15] is mainly composed of macrophages, microglia, antibodies, and complement. Clonally expanded B cells accumulate in MS lesions [16] as well as in the CSF. In particular, the preferential V(H) gene usage in the CSF and blood of MS patients was examined by PCR technologies, and analysis of HCDR3 length revealed an oligoclonal accumulation of B cells in the CSF. Sequence analysis of the V(H)3 and V(H)4 gamma transcripts of MS individuals demonstrated that this accumulation was related to the expansion and somatic diversification of a limited group of B-cell clones, indicative of a chronic and intense antigenic stimulation occurring inside the CNS [17, 18]. Moreover, when the CSF of MS patients was examined at different timepoints, a substantial proportion of shared clones in the samples taken at different times was found and these clones were identical or closely related; i.e., they had the same third complementary determining region (CDR) of the H chain variable region gene (HCDR3), with different mutations in the V(H) segment, suggestive of the presence of a continuous antigenic stimulation which persists with time in the CNS [19]. When subsets of B cells are examined in the CSF of MS patients, using flow cytometry, all the stages of B cell differentiation are observed, as well as the presence of lymphotoxin-α, CXCL12, and CXCL13, demonstrating that the process of B-cell maturation occurs inside the CNS, and in particular in the CSF, from where pathogenetic autoantibodies may attack the nervous tissue surrounding the ventricles [20]. It is of particular interest that newly formed lymphoid follicles have been recently detected in the meninges of MS patients with a secondary progressive clinical course; these ectopic B-cell follicles with germinal centers could represent a critical step in maintaining the compartmentalization of the disease and the chronic humoral autoimmune process [21]. All these data indicate that the B-cell response is of relevance in the pathogenesis of MS, and that in the future we will have to utilize, at least after the first stages of the disease, drugs which cross the blood-brain barrier, because the inflammatory process soon becomes sequestered inside the CNS and the CSF.

Cortical and Gray Matter Lesions

The presence of cortical lesions and the involvement of the gray matter were described by the first neuropathologists in the late 1800s, but in recent years these changes have been particularly well studied and evaluated, because it is now clear that brain atrophy, which occurs during the course of the disease and correlates with disability, is not only due to the loss of myelin tracts, but to a diffuse involvement of the cerebral cortex and deep gray matter, such as the basal ganglia, brainstem nuclei, and gray matter of the spinal cord. Considering that axons in the cortex are usually myelinated, demyelinated areas can also be expected to occur in the gray matter of the cortical regions. The demyelinating lesions can be detected in the deeper areas of the cortex, where myelination is more evident, but also in the subpial regions, or in the central part of the cortex [22]. It is interesting that, unlike in the white matter, the cortical areas of demyelination are not usually associated with a clear lymphocyte infiltration, thus suggesting that different mechanisms are probably involved [23]. Neuronal density can be significantly reduced, especially in secondary progressive MS and, while synaptic loss is not evident in some cases [23], it can be particularly striking in others [24]. Certain cortical areas, such as the cingulated gyrus, the insular cortex, and the temporal-basal cortex are usually more affected [25]. Cortical lesions can be relevant for the appearance, in some cases, of cognitive disturbances or other symptoms, such as seizures.

Normal-Appearing White Matter

Microglial nodules, small foci of inflammation or scattered axonal changes in the normal-appearing white matter (NAWM), were already known to the first neuropathologists, but renewed interest in the study of NAWM now comes from magnetic resonance imaging research, which has shown a diffuse injury of this area of the brain, which occurs early in the course of the disease and is usually also relevant in cases with a low total lesion load [26-29]. A recent study which specifically addressed this problem showed that in relapsing/remitting MS inflammation is mainly located in the demyelinated plaques, while in the progressive phase of the disease it is more diffuse in the whole of the white matter [30]. The inflammatory changes are characterized by perivascular foci of lymphocytes, diffuse infiltration of the NAWM by lymphocytes, microglial activation, meningeal inflammation, and axonal injury (Fig. 2). Using dysferlin expres-

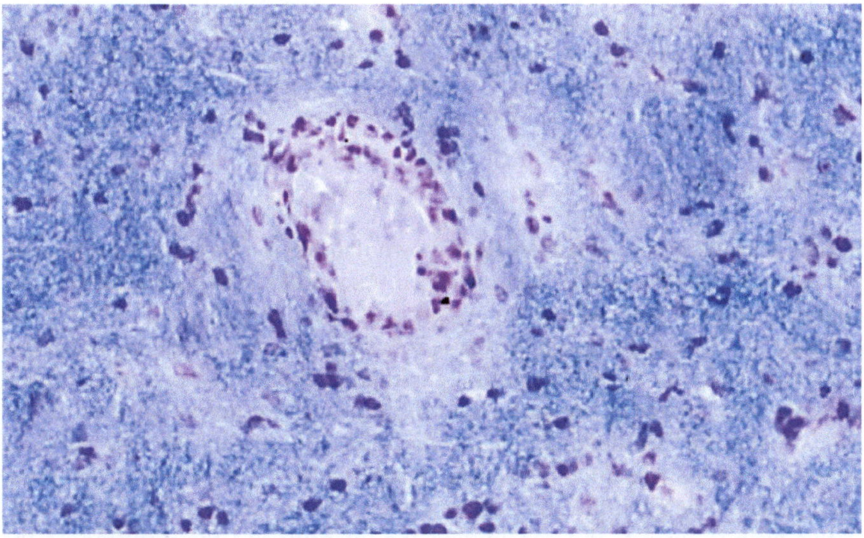

Fig. 2. Small perivascular inflammatory infiltrate in the normal-appearing white matter (H&E stained, 40×)

sion of leaky brain blood vessels, it was demonstrated that blood-brain barrier damage is not restricted to the demyelinating plaques but is diffuse outside of the lesion areas [31]. All these observations show that, after the first early phases, the inflammatory cells colonize the entire CNS, and the pathological process, which is diffuse and generalized, continues inside the brain and spinal cord where the majority of drugs that we now utilize cannot reach.

Oligodendrocyte Apoptosis and Microglial Activation

Recently, the presence of extensive oligodendrocyte apoptosis and microglial activation without lymphocytic infiltration or myelin phagocytes has been observed in some relapsing/remitting MS patients with a short disease duration [32]. Apoptosis of oligodendrocytes was previously described in MS, especially in the "type-3 lesion pattern" [15] but, according to the observations of Barnett and Prineas [33], this change is not limited to a subset of patients but more probably represents a general phenomenon of the early stage of plaque formation. According to these data, therefore, the damage to oligodendrocytes is not secondary to the inflammatory process mediated by autoreactive T and B lymphocytes, but is a primary event.

The role of microglia has been also particularly emphasized in recent years, not only as autoantigens presenting cells to autoreactive T cells or as cells secreting proinflammatory cytokines, but also as the primary cells which first react to the death of oligodendrocytes [34]. At present, it is not clear whether the apoptosis of oligodendrocytes and the activation of microglia is a primary general phenomenon or merely a pathological event which can occur during the formation of the demyelinating plaque.

Conclusions

Recent neuropathological studies have shown that MS is a very much more complex disease than previously acknowledged. Axonal damage occurs early in the course of disease, especially in the progressive phase, and B and T cells rapidly colonize the CNS and damage the tissue remaining trapped inside the brain and the spinal cord. The NAWM is diffusely injured by inflammatory and microglial cells, the gray matter in the cortex and in deep nuclei is deeply involved in the process, and the oligodendrocytes can die, also without showing clear signs of inflammation. At present, considering all the available data, the primary event still appears to be the infiltration of the CNS by autoreactive inflammatory cells. If this view is correct, and considering the intensity and the diffusion of the process, new therapeutic strategies need to be pursued urgently.

References

1. Lassmann H (2005) Multiple sclerosis pathology: evolution of pathogenetic concepts. Brain Pathol 15:217-22
2. Rivers TM, Sprunt DH, Berry GP (1933) Observations on attempts to produce acute disseminated encephalomyelitis in monkeys. J Exp Med 58:39-53
3. Patrikios P, Stadelmann C, Kutzelnigg A et al (2006) Remyelination is extensive in a subset of multiple sclerosis patients. Brain 129:3165-3172
4. Ferguson B, Matyszak MK, Esiri MM, Perry VH (1997) Axonal damage in acute multiple sclerosis lesions. Brain 120:393-399
5. Trapp BD, Peterson J, Ransohoff RM et al (1998) Axonal transection in the lesions of multiple sclerosis. N Engl J Med 338:278-285
6. Bitsch A, Schuchardt J, Bunkowski S et al (2000) Acute axonal injury in multiple sclerosis: correlation with demyelination and inflammation. Brain 123:1174-1183
7. Mancardi G, Hart B, Roccatagliata L et al (2001) Demyelination and axonal damage in a non-human primate model of multiple sclerosis. J Neurol Sci 184:41-49
8. Lovas G, Szilagyi N, Majtenyi K et al (2000) Axonal changes in chronic demyelinated cervical spinal cord plaques. Brain 123:308-317
9. Babbe H, Roers A, Waisman A et al (2000) Clonal expansions of CD8(+) T cells dominate the T cell infiltrate in active multiple sclerosis lesions as shown by micromanipulation and single cell polymerase chain reaction. J Exp Med 192:393-404
10. Bjartmar C, Yin X, Trapp BD (1999) Axonal pathology in myelin disorders. J Neurocytol 28:383-395
11. Nobbio L, Gherardi G, Vigo T et al (2006) Axonal damage and demyelination in long-term dorsal root ganglia cultures from a rat model of Charcot-Marie-Tooth type 1A disease. Eur J Neurosci 23:1445-1452
12. Coman I, Aigrot MS, Seilhean et al (2006) Nodal, paranodal and juxtaparanodal axonal proteins during demyelination and remyelination in multiple sclerosis. Brain 129:3186-3195

13. Genain CP, Cannella B, Hauser SL, Raine CS (1999) Identification of autoantibodies associated with myelin damage in multiple sclerosis. Nat Med 5:170-175
14. Berger T, Rubner P, Schautzer F et al (2003) Antimyelin antibodies as a predictor of clinically definite multiple sclerosis after a first demyelinating event. N Engl J Med 349:139-145
15. Lucchinetti C, Bruck W, Parisi J et al (2000) Heterogeneity of multiple sclerosis lesions: implications for the pathogenesis of demyelination. Ann Neurol 47:707-717
16. Baranzini SE, Jeong MC, Butunoi C et al (1999) B cell repertoire diversity and clonal expansion in multiple sclerosis brain lesions. J Immunol 163:5133-5144
17. Colombo M, Dono M, Gazzola P et al (2000) Accumulation of clonally related B lymphocytes in the cerebrospinal fluid of multiple sclerosis patients. J Immunol 164:2782-2789
18. Mancardi G, Hart BA, Capello E et al (2000) Restricted immune responses lead to CNS demyelination and axonal damage. J Neuroimmunol 107:178-183
19. Colombo M, Dono M, Gazzola P et al (2003) Maintenance of B lymphocyte-related clones in the cerebrospinal fluid of multiple sclerosis patients. Eur J Immunol 33:3433-3438
20. Corcione A, Casazza S, Ferretti E et al (2004) Recapitulation of B cell differentiation in the central nervous system of patients with multiple sclerosis. Proc Natl Acad Sci USA 101:11064-11069
21. Serafini B, Rosicarelli B, Magliozzi R et al (2006) Dendritic cells in multiple sclerosis lesions: maturation stage, myelin uptake, and interaction with proliferating T cells. J Neuropathol Exp Neurol 65:124-141
22. Bo L, Vedeler CA, Nyland HI et al (2003) Subpial demyelination in the cerebral cortex of multiple sclerosis patients. J Neuropathol Exp Neurol 62:723-732
23. Vercellino M, Plano F, Votta B et al (2005) Grey matter pathology in multiple sclerosis. J Neuropathol Exp Neurol 64:1101-1107
24. Wegner C, Esiri MM, Chance SA et al (2006) Neocortical neuronal, synaptic, and glial loss in multiple sclerosis. Neurology 67:960-967
25. Kutzelnigg A, Lassmann H (2006) Cortical demyelination in multiple sclerosis: a substrate for cognitive deficits? J Neurol Sci 245:123-126
26. De Stefano N, Narayanan S, Francis SJ et al (2002) Diffuse axonal and tissue injury in patients with multiple sclerosis with low cerebral lesion load and no disability. Arch Neurol 59:1565-1571
27. Ciccarelli O, Werring DJ, Wheeler-Kingshott CA et al (2001) Investigation of MS normal-appearing brain using diffusion tensor MRI with clinical correlations. Neurology 56:926-933
28. Pelletier D, Nelson SJ, Oh J et al (2003) MRI lesion volume heterogeneity in primary progressive MS in relation with axonal damage and brain atrophy. J Neurol Neurosurg Psychiat 74:950-952
29. Rocca MA, Iannucci G, Rovaris M et al (2003) Occult tissue damage in patients with primary progressive multiple sclerosis is independent of T2-visible lesions: a diffusion tensor MR study. J Neurol 250:456-460
30. Kutzelnigg A, Lucchinetti CF, Stadelmann C et al (2005) Cortical demyelination and diffuse white matter injury in multiple sclerosis. Brain 128:2705-2712
31. Hochmeister S, Grundtner R, Bauer J et al (2006) Dysferlin is a new marker for leaky brain blood vessels in multiple sclerosis. J Neuropathol Exp Neurol 65:855-865
32. Matute C, Perez-Cerda F (2005) Multiple sclerosis: novel perspectives on newly forming lesions. Trends Neurosci 28:173-175
33. Barnett MH, Prineas JW (2004) Relapsing and remitting multiple sclerosis: pathology of the newly forming lesion. Ann Neurol 55:458-468
34. Deng X, Sriram S (2005) Role of microglia in multiple sclerosis. Curr Neurol Neurosci Rep 5:239-244

Chapter 2

Neurophysiology

L. Leocani, G. Comi

Introduction

Neurophysiological methods, particularly evoked potentials (EPs; also known as evoked responses), are widely applied in the functional assessment of multiple sclerosis (MS), since they provide a quite reliable, even though indirect, measure of the extent of demyelination or axonal loss in a given nerve pathway. For this reason, they are used to indicate the involvement of sensory and motor pathways in the presence of vague disturbances and to detect clinically silent lesions, even though the latter application has been greatly reduced since the development of magnetic resonance imaging (MRI), which is more sensitive in detecting subclinical lesions. Nevertheless, the information provided by EPs is more strictly related to function than is the information obtained from structural MRI techniques. As O'Connor et al. [1] point out, it is impossible to "confidentially predict, from examining an MS patient's cranial MRI, what the clinical findings or EDSS score will be." In fact, the severity of the disease, assessed clinically, correlates well with the degree of neurophysiological abnormality found [2-4]. We briefly review the application of EPs in the assessment of the pathophysiology, diagnosis, and monitoring of MS.

Diagnosis and Pathophysiology

The pathological substrates of EP abnormalities in MS are demyelination and axonal loss [5-7]. In myelinated fibers, saltatory conduction of action potentials is determined by clustering of voltage-sensitive sodium channels within axon membranes at nodes of Ranvier and, to a much lesser extent, beneath the myelin sheath [8]. Demyelination may produce conduction block [9-11], which is also produced by soluble mediators of inflammation [12-14].

In areas with partial demyelination, slowing of conduction velocity and a prolonged refractory period with failure in transmitting high-frequency impulses may occur [10]. As a result, EP abnormalities may consist of delayed latency, morphological abnormalities, wave cancellation, amplitude reduction, and increased refractory period [15, 16]. Axonal loss, which is also an important feature in MS, especially in the progressive phase [17], also contributes to EP abnormalities [18]. Even though EP abnormalities may reveal subclinical lesions [16, 19], their value in the diagnosis of definite MS still needs to be clarified [20], even though it is much lower than that of MRI due to its lower sensitivity [16]. Previous studies [21, 22] have indicated that in isolated syndromes about one-third of the patients have subclinical involvement of sensory pathways revealed by EPs, mostly by visual and somatosensory evoked potentials. In a recent study on 112 patients with isolated optic neuritis [23], 34.1% of patients had abnormal extravisual EPs; however, the contribution of neurophysiological techniques in demonstrating spatial dissemination of the lesions was quite poor: only 4% of patients with abnormal extra-visual EPs had normal brain MRI. The major limiting factor on the usefulness of EPs in detecting subclinical involvement is that the presence of a lesion is revealed only if it affects pathways explored by neurophysiological investigations; moreover, a significant proportion of the fibers must be affected to produce recordable modifications of evoked responses [24]. Stimulus manipulation [25] or more sophisticated recording techniques [26] or a combination of the two [27] may improve sensitivity in the assessment of nervous function, but raises the issue of standardization among different laboratories. Therefore, although the International Panel criteria for MS diagnosis [28], largely based on MRI findings, have been criticized [29], only visual EPs are viewed as contributing to the diagnosis of MS [28].

Assessment of Disease Severity

Apart from MS diagnosis, the use of EPs seems more promising in the assessment and monitoring of disease severity, in relation to their good correspondence with function of the investigated pathway. In fact, EPs have been reported to correlate with clinical findings in somatosensory [30, 31], visual [32], and motor pathways [33]. Moreover, several studies have reported that the combination of EP abnormalities correlates with disability [34].

In a study of 40 patients with relapsing-remitting (RR) and 13 with secondary progressive (SP) MS [35], spinal motor conduction time and the frequency of abnormalities of multimodal EPs were significantly greater in SPMS patients than in controls and RRMS patients. Spinal motor conduction times also correlated directly with scores on the Expanded Disability Status Scale (EDSS) and pyramidal functional system (FS) scores of Kurtzke [36]; while brain lesion load, evaluated in SPMS, did not correlate with disability scores. The authors suggest that disability in SPMS patients is mainly due to progressive involvement of the corticospinal tract in the spinal cord. Filippi et al. [3] compared brain MRI and multimodal EPs in patients with benign and SPMS, EP abnormalities were significantly more frequent and more severe in the latter group, according to their higher disability. Similarly, Kira et al. [37] found a significantly higher frequency of abnormal records in visual, brainstem auditory, and somatosensory EPs in primary progressive (PP) MS patients than in

RRMS patients. Moreover, clinically unexpected abnormalities were significantly more common in PPMS than in RRMS patients throughout all EP modalities.

Monitoring of Disease Evolution

Sater et al. [38] performed a longitudinal study (mean follow-up 1.5 years) of BAEP and VEP in a group of 11 chronic progressive MS patients: P100 latency significantly increased, while the BAEP I-V interpeak latency, T2 lesion load, and EDSS score did not. These data suggest that VEPs may be more sensitive than clinical and conventional brain MRI measures in detecting disease evolution in the progressive phase, although they need to be validated with a larger group sample. Kidd et al. [39] evaluated longitudinally central motor conduction time and brain and cervical MRI by means of transcranial magnetic stimulation in a group of 10 PPMS and 10 SPMS patients. Central motor conduction time was weakly, but significantly, correlated with EDSS. After 1 year, the increase in central motor conduction time was not significantly correlated with changes in cord lesions or area, but was present only in the four patients with an increased number of cord lesions. Fuhr et al. [34] found cross-sectional and longitudinal correlation between combined VEP and MEP latency *z*-score and EDSS in a group of 30 MS patients (the majority with RRMS) who had been followed for 2 years. The long-term (up to 3 years) follow-up of VEP in patients with acute optic neuritis revealed a significant decrease in VEP latency in the affected eye; on the other hand, the VEP latency increased in the unaffected eye, suggesting the possible role of remyelination (in the affected eye) and of the subclinical disease activity [40]. In a 1-year follow-up study in 90 patients with optic neuritis (monosymptomatic in 58 patients and part of clinically definite MS in 32), the mean latency of VEPs in eyes with optic neuritis was longer in the clinically definite MS patients [41]. Significant effects of time after onset were observed for latencies, amplitudes, and VEP abnormality scores.

For the correlation between clinical and EP findings, conventional scores could provide some advantage with respect to parametric values such as latency and amplitude, since the latter may be more influenced by test-retest variability, especially in MS patients [42], and do not allow for the consideration of absent components. The Evoked Potentials Abnormality Score (EPAS) has been used together with conventional brain MRI and EDSS to evaluate longitudinal changes in a 2-year follow-up study performed in 50 MS patients [1]. While MRI was significantly correlated with EDSS only at the cross-sectional evaluation at year 1, the EPAS was significantly correlated with EDSS and MRI at the cross-sectional evaluations, and only with EDSS at the longitudinal evaluation. A conventional score has been applied in another longitudinal study on 84 MS patients with the RR or progressive form of the disease [4]. In agreement with previous literature [43-46], visual evoked potentials (VEP), lower limb sensory evoked potentials (SEP) and motor evoked potentials (MEP) were the EPs most frequently involved in MS.

These findings may be explained by a higher susceptibility of optic nerve fibers to MS lesions, confirmed by pathological studies [47], and by a higher probability of involvement of longer pathways such as sensorimotor projections to the lower limbs. In addition, abnormalities of EPs were more frequent and severe in progressive forms of the disease compared with the RR form. Cross-sectionally, the severity of each EP score significantly correlated with the corresponding FS for all but follow-up visual EP, and with EDSS for all but brainstem EPs. EDSS significantly correlated with global EP score severity. Longitudinal EP/FS correlations were significant only in the somatosensory system. However, MS patients with disability progression at follow-up had more severe baseline EP scores than patients who remained stable. Patients with a severe baseline global EP score had a higher risk (72%) of disability progression at follow-up compared with MS patients with a lower score (36.3%). The predictive value of EPs has previously been pointed out using combined visual and motor EP latencies [39], and using sum scores for EP abnormalities [48]. In the latter study, EPs were retrospectively examined in 94 MS patients at first presentation and after 5 and 10 years. In patients examined early after disease onset, a significant predictive value for abnormal EP was found with MEP, SEP (but not VEP), and cumulative sum scores at first presentation, associated with higher EDSS values after 5 years.

There have previously been several attempts to use EPs in clinical trials [16, 17]. Short-term (2 and 26 weeks) effects of intravenous methylprednisolone on VEP in patients with acute optic neuritis have been investigated [49]. After 2 and 26 weeks, both amplitude and latency were not significant in patients treated with steroids and in patients who received a placebo. Interestingly enough, no treatment effects were observed on the length of the optic nerve lesions. Sheean et al. [50], in an open-label study in MS patients complaining of fatigue, did not find significant changes in MEP parameters in patients treated with 3,4-diaminopyridine compared with untreated patients, while a rectified electromyogram (EMG) did show a treatment effect. The results of this study indicate that the MEP parameters utilized (central motor conduction time and MEP size) are not useful in evaluating the effects of drugs on fatigue, a result explained by the absence of effects of exercise on the same parameters. It has to be considered that the study by Sheean et al. [50] was performed on the upper limb, which was relatively unaffected by the disease. In fact, a previous study had reported that walking reduced MEP area more in MS patients complaining of fatigue than in normal controls [51]. After intrathecal baclofen treatment for spasticity on 11 MS patients, upper-limb MEP amplitude and area revealed changes in corticospinal output, while MEP threshold and central motor conduction times remained unchanged [52]. By analyzing MEP changes in response to a fatigue paradigm (3 min of maximal contraction) in eight MS patients complaining of fatigue, a significant reduction in MEP depression following 6 months of IFNb treatment, together with a reduction in subjective physical fatigue, was demonstrated [53]. Again, in this study, the MEP latency did not significantly change over time. Using the triple stimulation technique (TST), to allow a more accurate quantification of conducting corticospinal neurons compared with single stimulation, Humm et al. [54] reported an increased TST amplitude ratio in patients with RRMS and SPMS patients after treatment of acute exacerbations with methylprednisolone, but not in PPMS patients undergoing the same treatment; this is consistent with the limited clinical efficacy of this treatment in the latter group.

Other Neurophysiological Methods

Standard EPs evaluate the main sensory and motor projection pathways, but do not provide information about the associative and commissural pathways, constituting the great majority of the brain white matter, whose involvement is considered the major determinant in the pathophysiology of cognitive impairment in MS. Cognitive function is impaired in about 50% of MS patients [55]. The pattern of cognitive dysfunction is typical of subcortical dementia [56, 57], explained by the disconnection of large portions of the cortical associative areas, occurring as a consequence of demyelination and axonal degeneration [57-59]. Moreover, the greater importance of lesions immediately underlying the cortex, with respect to other lesion locations, has been proposed [60, 61]. The electroencephalogram (EEG), which is the expression of multiple neuronal network interactions affected by white matter damage, may be used as an indicator of the global status of such interactions [17]. Spectral analysis of the EEG revealed abnormalities in 40-79% of MS patients [62, 63], mainly an increase in slow frequency activity and a decrease in alpha band activity. In progressive MS patients, these abnormalities have been found to relate to cognitive dysfunction [19]. In that study, EEG spectral power and coherence were examined in a group of 28 progressive MS patients with or without cognitive impairment assessed by a battery of neuropsychological tests. Cognitively impaired MS patients had a significant increase in theta power over the frontal regions and a diffuse coherence decrease, not found in cognitively intact patients. Moreover, coherence decrease, indicating involvement of functional cortico-cortical connections [64], was significantly correlated with brain MRI lesion load immediately underlying the cortex. These findings are consistent with other MRI studies, demonstrating a statistically significant correlation between global cognitive impairment and the severity of white matter abnormalities of the hemispheres, of corpus callosum atrophy, and of ventricular enlargement [16, 65-67]. A closer correlation has been found between cognitive impairment and both subcortical lesion load [60, 61] and corpus callosum atrophy [68]. Moreover, the analysis of regional cerebral lesion load showed significant relationships with specific cognitive functions [66-68]. Another neurophysiological test for the investigation of brain function is the analysis of event-related potentials (ERPs). The most widely studied ERP, P300, is a positive wave recorded over the scalp when subjects discriminate between stimuli differing in some physical dimension. It is thought to represent a closure of the process of stimulus evaluation [69], whose latency has been proposed as an indicator of information processing speed [70]. This process is electively affected in MS [55], and in fact P300 latency is increased in MS patients [71, 72]. Even though the value of P300 in the classification of single patients is limited, the increase in P300 latency is correlated with the severity of cognitive impairment [73, 74] and, interestingly, with the degree of white matter involvement [71, 74]. Also the mismatch negativity, a pre-attentive component preceding P300, has been reported as being of reduced amplitude in MS patients, particularly in those with cognitive impairment [75]. Moreover, ERPs to specific cognitive tasks may help in assessing specific cognitive domains, such as abstract reasoning and memory, which are electively impaired in MS [68]. Pelosi et al. [77] investigated visual and auditory ERPs elicited in the Sternberg paradigm [78], assessing working memory, on a group of patients with clinically isolated myelopathy suggestive of MS. The component of the response that has been shown to be sensitive to memory loading in healthy control subjects was affected in patients with memory dysfunction. ERPs have been used for monitoring MS evolution and the effects of treatment. In a double-blind study design, the effects of a high dose of intravenous methylprednisolone were investigated in 44 patients with clinically active MS [79]. The latency of P300 was significantly shortened after treatment but not after placebo. These findings, if confirmed by further studies, may elucidate the role to ERPs in monitoring cognitive function in clinical trials.

Conclusions

Standard evoked potentials are easily performed in most neurophysiology laboratories, with only minor discomfort for the patients. In the past, their application in MS has been limited to diagnostic purposes, but their usefulness in this direction is small when compared with the more sensitive MRI techniques. A promising application of EPs seems to be the assessment of disease severity for the detection of worsening, and possibly for monitoring the natural evolution of the disease and the effect of therapeutic interventions, due to the good correlation between EP abnormalities and disability. The application of neurophysiological techniques in providing a measure of the overall brain involvement, such as EEG and ERPs, may represent a useful tool for the assessment and monitoring of cognitive functions and may provide information on the physiopathology of the disease.

References

1. O'Connor P, Marchetti P, Lee L, Perera M (1998) Evoked potential abnormality scores are a useful measure of disease burden in relapsing-remitting multiple sclerosis. Ann Neurol 44:404-407

2. Nuwer MR, Packwood JW, Myers LW et al (1987) Evoked potentials predict the clinical changes in a multiple sclerosis drug study. Neurology 37:1754-1761
3. Filippi M, Campi A, Mammi S et al (1995) Brain magnetic resonance imaging and multimodal evoked potentials in benign and secondary progressive multiple sclerosis. J Neurol Neurosurg Psychiatr 58:31-37
4. Leocani L, Rovaris M, Boneschi FM et al (2006) Multimodal evoked potentials to assess the evolution of multiple sclerosis: a longitudinal study. J Neurol Neurosurg Psychiatry 77:1030-1035
5. Lassmann H, Wisniewski HM (1979) Chronic relapsing experimental allergic encephalomyelitis: clinicopathological comparison with multiple sclerosis. Arch Neurol 36:490-497
6. Trapp BD, Peterson J, Ransohoff RM et al (1998) Axonal transection in the lesions of multiple sclerosis. N Engl J Med 338:278-285
7. Scolding N, Franklin R (1998) Axon loss in multiple sclerosis. Lancet 352:340-341
8. Ritchie JM, Rogart RB (1977) Density of sodium channels in mammalian myelinated nerve fibers and nature of the axonal membrane under the myelin sheath. Proc Natl Acad Sci USA 74:211-215
9. McDonald WI (1963) The effects of experimental demyelination on conduction in peripheral nerve: a histological and electrophysiological study. II. Electrophysiological observations. Brain 86:501-524
10. McDonald WI, Sears TA (1970) The effects of experimental demyelination on conduction in the central nervous system. Brain 93:583-598
11. Rasminsky M, Sears TA (1972) Internodal conduction in undissected demyelinated fibres. J Physiol (Lond) 227:323-50
12. Moreau T, Coles A, Wing M et al (1996) Transient increase in symptoms associated with cytokine release in patients with multiple sclerosis. Brain 119:225-237
13. Koller H, Siebler M, Hartung HP (1997) Immunologically induced electrophysiological dysfunction: implications for inflammatory diseases of the CNS and PNS. Prog Neurobiol 52:1-26
14. Smith KJ, Lassmann H (2002) The role of nitric oxide in multiple sclerosis. Lancet Neurol 1:232-41
15. Emerson RG (1998) Evoked potentials in clinical trials for multiple sclerosis. J Clin Neurophysiol 15:109-116
16. Comi G, Locatelli T, Leocani L et al (1999) Can evoked potentials be useful in monitoring multiple sclerosis evolution? Electroencephalogr Clin Neurophysiol Suppl 50:349-357
17. Bjartmar C, Trapp BD (2001) Axonal and neuronal degeneration in multiple sclerosis: mechanisms and functional consequences. Curr Opin Neurol 14:271-278
18. McGavern DB, Murray PD, Rivera-Quinones C et al (2000) Axonal loss results in spinal cord atrophy, electrophysiological abnormalities and neurological deficits following demyelination in a chronic inflammatory model of multiple sclerosis. Brain 3:519-531
19. Leocani L, Comi G (2000) Neurophysiological investigations in multiple sclerosis. Curr Opin Neurol 13:255-261
20. Gronseth GS, Ashman EJ (2000) Practice parameter: the usefulness of evoked potentials in identifying clinically silent lesions in patients with suspected multiple sclerosis (an evidence-based review): Report of the Quality Standards Subcommittee of the American Academy of Neurology. Neurology 54:1720-1725
21. Filippini G, Comi G, Cosi V et al (1994) Sensitivities and predictive values of paraclinical tests for diagnosing multiple sclerosis. J Neurol 241:132-137
22. Frederiksen JL, Petrera J, Larsson HBW et al (1996) Serial MRI, VEP, SEP, and biotesiometry in acute optic neuritis: value of baseline results to predict the development of new lesions at one-year follow-up. Acta Neurol Scand 93:246-252
23. Ghezzi A, Martinelli V, Torri V et al (1999) Long-term follow-up of isolated optic neuritis: the risk of developing multiple sclerosis, its outcome, and the prognostic role of paraclinical tests. J Neurol 246:770-775

24. Comi G, Leocani L, Medaglini S et al (1999) Measuring evoked responses in multiple sclerosis. Mult Scler 5:263-267
25. Parisi V, Pierelli F, Restuccia R et al (1998) Impaired VEP after photostress response in multiple sclerosis patients previously affected by optic neuritis. Electroencephalogr Clin Neurophysiol 108:73-79
26. Onofrj M, Bazzano S, Malatesta G, Gambi D (1990) Pathophysiology of delayed evoked potentials in multiple sclerosis. Funct Neurol 5:301-319
27. Humm AM, Beer S, Kool J et al (2004) Quantification of Uhthoff's phenomenon in multiple sclerosis: a magnetic stimulation study. Clin Neurophysiol 115:2493-2501
28. McDonald W, Compston D, Edan G et al (2001) Recommended diagnostic criteria for multiple sclerosis: guidelines from the international panel on the diagnosis of multiple sclerosis. Ann Neurol 50:121-127
29. Poser CM (2005) The diagnosis and management of multiple sclerosis. Acta Neurol Scand 112:199-201
30. Leocani L, Martinelli V, Natali-Sora MG et al (2003) Somatosensory evoked potentials and sensory involvement in multiple sclerosis: comparison with clinical findings and quantitative sensory tests. Mult Scler 9:275-279
31. Fukutake T, Kuwabara S, Kaneko M et al (1998) Sensory impairments in spinal multiple sclerosis: a combined clinical, magnetic resonance imaging and somatosensory evoked potential study. Clin Neurol Neurosurg 100:199-204
32. Weinstock-Guttman B, Baier M, Stockton R et al (2003) Pattern reversal visual evoked potentials as a measure of visual pathway pathology in multiple sclerosis. Mult Scler 9:529-534
33. Van der Kamp W, Maertens de Noordhout A, Thompson PD et al (1991) Correlation of phasic muscle strength and corticomotoneuron conduction time in multiple sclerosis. Ann Neurol 29:6-12
34. Fuhr P, Borggrefe-Chappuis A, Schindler C et al (2001) Visual and motor evoked potentials in the course of multiple sclerosis. Brain 124:2162-2168
35. Facchetti D, Mai R, Colombo A et al (1994) Limited clinical significance of traditional and quantitative EEG in multiple sclerosis. Acta Neurol Belg 94:245-250
36. Kurtzke JF (1983) Rating neurological impairment in multiple sclerosis: an Expanded Disability Status Scale (EDSS). Neurology 33:1444-1452
37. Kira J, Tobimatsu S, Goto I, Hasuo K (1993) Primary progressive versus relapsing remitting multiple sclerosis in Japanese patients: a combined clinical, magnetic resonance imaging and multimodality evoked potential study. J Neurol Sci 117:179-185
38. Sater RA, Rostami AM, Galetta S et al (1999) Serial evoked potential studies and MRI imaging in chronic progressive multiple sclerosis. J Neurol Sci 171:79-83
39. Kidd D, Thompson PD, Day BL et al (1998) Central motor conduction time in progressive multiple sclerosis: correlations with MRI and disease activity. Brain 121:1109-1116
40. Brusa A, Jones SJ, Kapoor R et al (1999) Long-term recovery and fellow eye deterioration after optic neuritis, determined by serial visual evoked potentials. J Neurol 246:776-782
41. Frederiksen JL, Petrera J (1999) Serial visual evoked potential in 90 untreated patients with acute optic neuritis. Surv Ophthalmol 44:S54-S62
42. Andersson T, Persson A (1990) Reproducibility of somatosensory evoked potentials (SEPs) after median nerve stimulation. Electroencephalogr Clin Neurophysiol 30:205-211
43. Chiappa KH (1980) Pattern shift visual, brainstem auditory and short latency somatosensory evoked potentials in multiple sclerosis. Neurology 30:110-123
44. Khosbin S, Hallett M (1981) Multimodality evoked potentials and blink reflex in multiple sclerosis. Neurology 31:138-144

45. Trojaborg W, Petersen E (1979) Visual and somatosensory evoked potentials in multiple sclerosis. J Neurol Neurosurg Psychiat 42:323-330
46. Comi G, Martinelli V, Medaglini S et al (1989) Correlation between multimodal evoked potentials and magnetic resonance imaging in multiple sclerosis. J Neurol 236:4-8
47. Dawson JW (1916) The histology of disseminated sclerosis. Edinburgh Med J (NS) 17:311-410
48. Kallmann BA, Fackelmann S, Toyka KV et al (2006) Early abnormalities of evoked potentials and future disability in patients with multiple sclerosis. Mult Scler 12:58-65
49. Kapoor R, Miller DH, Jones SJ et al (1998) Effects of intravenous methylprednisolone on outcome in MRI-based prognostic subgroups in acute optic neuritis. Neurology 50:230-237
50. Sheean GJ, Murray NM, Rothwell JC et al (1998) An open-labelled clinical and electrophysiological study of 3,4-diaminopyridine in the treatment of fatigue in multiple sclerosis. Brain 121:967-975
51. Schubert M, Wohlfart K, Rollnik JD, Dengler R (1998) Walking and fatigue in multiple sclerosis: the role of the corticospinal system. Muscle Nerve 21:1068-1070
52. Auer C, Siebner HR, Dressnandt J, Conrad B (1999) Intrathecal baclofen increases corticospinal output to hand muscles in multiple sclerosis. Neurology 52:1298-1299
53. White AT, Petajan JH (2004) Physiological measures of therapeutic response to interferon beta-1a treatment in remitting-relapsing MS. Clin Neurophysiol 115:2364-2371
54. Humm AM, Z'Graggen WJ, Buhler R et al (2006) Quantification of central motor conduction deficits in multiple sclerosis patients before and after treatment of acute exacerbation by methylprednisolone. J Neurol Neurosurg Psychiat 77:345-350
55. Rao SM, Leo GJ, Bernardin L, Unverzagt (1991) Cognitive dysfunction in multiple sclerosis. I. Frequency, patterns and predictions. Neurology 41:685-691
56. Filley CM, Heaton RK, Nelson LM et al (1989) A comparison of dementia in Alzheimer's disease and multiple sclerosis. Arch Neurol 46:157-161
57. Comi G, Filippi M, Martinelli V et al (1993) Brain magnetic resonance imaging correlates of cognitive impairment in multiple sclerosis. J Neurol Sci 115:S66-S73
58. Rao SM (1990) Multiple sclerosis. In: Cummings JL (ed) Subcortical dementia. Oxford University Press, New York, pp 164-180
59. Mahler ME, Benson DF (1990) Cognitive dysfunction in multiple sclerosis: a subcortical dementia? In: Rao SM (ed) Neurobehavioral aspects of multiple sclerosis. Oxford University Press, New York, pp 88-101
60. Damian MS, Schilling G, Bachmann G et al (1994) White matter lesions and cognitive deficits: relevance of lesion pattern? Acta Neurol Scand 90:430-436
61. Miki Y, Grossman RI, Udupa JK et al (1998) Isolated U-fiber involvement in MS: preliminary observations. Neurology 50:1301-1306
62. Locatelli T, Filippi M, Martinelli V et al (1993) EEG mapping in multiple sclerosis. Riv Neurobiol 39:233-237
63. Thatcher RW, Krause PJ, Hrybyk M (1986) Cortico-cortical associations and EEG coherence: a two compartmental model. Electroencephalogr Clin Neurophysiol 64:123-143
64. Rao SM, Leo GJ, St Aubin-Faubert P (1989) On the nature of memory disturbance in multiple sclerosis. J Clin Exp Neuropsychol 11:699-712
65. Franklin GM, Nelson LM, Filter CM, Heaton RK (1989) Cognitive loss in multiple sclerosis. Arch Neurol 46:162-167
66. Swirsky-Sacchetti T, Mitchell DR, Seward J et al (1992) Neuropsychological and structural brain lesions in multiple sclerosis: a regional analysis. Neurology 42:1291-1295
67. Foong J, Rozewicz L, Quaghebeur G et al (1997) Executive functions in multiple sclerosis: the role of frontal lobe pathology. Brain 120:15-16

68. Rovaris M, Filippi M, Falautano M et al (1998) Relationship between MR abnormalities and patterns of cognitive impairment in multiple sclerosis. Neurology 50:1601-1608
69. Desmedt JE (1980) P300 in serial tasks: an essential post-decision closure mechanism. Progr Brain Res 54:682-686
70. Kutas M, McCarthy G, Donchin E (1977) Augmenting mental chronometry: the P300 as a measure of stimulus evaluation time. Science 197:792-795
71. Honig LS, Ramsay RE, Sheremata WA (1992) Event-related potentials P300 in multiple sclerosis: relation to magnetic resonance imaging and cognitive impairment. Arch Neurol 49:44-50
72. Newton MR, Barrett G, Callanan MM, Towell AD (1989) ERP P300 in multiple sclerosis. Brain 112:1636-1660
73. Giesser BS, Schroeder MM, La Rocca NG et al (1992) Endogenous event-related potentials as indices of dementia in multiple sclerosis patients. Electroencephalogr Clin Neurophysiol 82:320-329
74. Piras MR, Magnano I, Canu ED et al (2003) Longitudinal study of cognitive dysfunction in multiple sclerosis: neuropsychological, neuroradiological, and neurophysiological findings. J Neurol Neurosurg Psychiatry 74:878-885
75. Jung J, Morlet D, Mercier B et al (2006) Mismatch negativity (MMN) in multiple sclerosis: an event-related potentials study in 46 patients. Clin Neurophysiol. 117:85-93
76. Pelosi L, Geesken JM, Holly M et al (1997) Working memory impairment in early multiple sclerosis: evidence from an event-related potential study of patients with clinically isolated myelopathy. Brain 120:2039-2058
77. Sternberg S (1966) High-speed scanning in human memory. Science 153:652-654
78. Filipovic SR, Drulovic J, Stojsavlievic N, Levic Z (1997) The effects of high-dose intravenous methylprednisolone on event-related potentials in patients with multiple sclerosis. J Neurol Sci 152:147-153

MRI TECHNIQUES TO ASSESS NEURODEGENERATION

Chapter 3

Atrophy

W. Rashid, D.T. Chard, D.H. Miller

Introduction

In this chapter we provide a brief overview of atrophy measurements in people with multiple sclerosis (MS). In particular we highlight the main findings from previous MRI studies, the main methodologies used to measure atrophy including their pros and cons, and the clinical relevance of such measures.

As previous chapters have highlighted, neurodegeneration is a pathologically significant process in MS, and is seen in both normal-appearing and lesional tissues. When considering how magnetic resonance imaging (MRI) assesses neurodegeneration, it is important to remember that MRI achieves a resolution far short of that necessary to identify individual cells and, as such, MRI estimates parameters from relatively large and mixed cell populations, and that atrophy or loss of one cell type may be masked by hypertrophy or proliferation of another, or by other factors such as tissue edema. Indeed, previous work has suggested that the gray matter (GM) lesions in MS are less inflammatory than those in white matter (WM) [1], with limited lymphocytic infiltration [2] and complementary activation [3], and so it may be expected that WM volumes will be influenced more by inflammatory noise than will GM. Given that atrophy measures, as with all the MRI parameters in current use, cannot be said to be purely assessing neurodegeneration, it may be necessary to assess both intrinsic tissue characteristics – such as magnetization transfer or metabolite concentrations – along with measurements of tissue volume, in order to obtain accurate estimates of neurodegeneration.

Brain Atrophy

Previous studies in MS have demonstrated excess brain atrophy from the onset of the disease [4-6] through to secondary progressive (SP) disease [7-13]. Differences in the degree of atrophy may be observed in different disease phenotypes, although this is not necessarily consistent and may vary according to the technique employed. Pagani et al. [12] observed that in people with relapsing/remitting (RR) MS, compared with those with progressive disease, ventricular enlargement was more prominent than cortical atrophy, while in those with progressive disease, cortical atrophy was more noticeable, to a degree mirroring the findings of Kalkers et al. [10]. Tedeschi et al. [13] noted greater WM and GM atrophy in people with progressive MS versus RRMS, while Turner et al. [11] did not. Relatively few studies have assessed GM and WM atrophy simultaneously, but from those that have, it would appear that while WM atrophy may be greater than that of GM initially [14], early in the course of the disease GM atrophy starts to evolve more rapidly than that of WM [6, 15, 16].

Gray Matter Atrophy

Studies assessing cortical regional atrophy have not been entirely consistent, but have tended to find a degree of preferential frontal atrophy [17-20], and temporal involvement [18, 20]. Deep GM structures (thalami) also appear to be usually affected [17, 21]) even early in the clinical disease course [18].

Ventricular Volumes

Brain atrophy may also be assessed indirectly by measuring ventricular volumes, with cerebrospinal fluid (CSF) spaces directly reciprocating brain tissue volumes within the fixed volume of the skull. Significant ventricular enlargement has been observed in MS [4, 5, 10, 22-25], although from this it is not possible to determine whether this represents global or regional brain atrophy.

Brain Atrophy and Lesion Loads

Brain lesion loads and atrophy appear to be related only to a relatively modest degree, with some studies finding an association [12, 14, 17, 26-30], and others not [31-35]. This suggests that an element of atrophy may either be attributable directly to volume loss in lesions, or to volume loss in neural networks interrupted by them, but that a significant element, perhaps the majority, cannot be explained in this way. The relationship between lesions and atrophy may be complicated by a delay between the inflammatory event and the subsequent development of atrophy, and some studies have found an association between early lesion loads and later atrophy [36-38].

Brain Atrophy, Clinical Impairment, and Disability

Brain atrophy, as seen on MRI, does appear to be clinically relevant, correlating significantly with measures of clinical disability, although tending to explain only about a quarter of variability in clinical scores [7, 8, 11, 13, 17, 25, 26, 30, 32, 39-49]. The relatively limited magnitude of correlations may be an underestimate of the true degree of association for several reasons. Firstly, the precision of both MRI and clinical measurements is not perfect, and this error will limit the maximum observable correlation. Secondly, MRI measures of global atrophy may mask clinically relevant tissue-specific or regional atrophy, and in this regard the observations that GM atrophy may be more dynamic than WM atrophy, WM volumes may be more affected by inflammatory noise, and atrophy may show regional variability within a given tissue type, may all be relevant. Thirdly, commonly employed clinical measures, such as the Expanded Disability Status Scale (EDSS) [50], are effectively composites of multiple functions, each of which may be affected by pathology in a variety of locations within the central nervous system (CNS).

Cord Atrophy

Neurodegeneration is likely to be regionally specific, as a number of studies have shown that there is little or no relationship between brain and spinal cord tissue loss as estimated by atrophy measures in MS [26, 51-53]. This suggests that the mechanisms responsible for atrophy in these two components of the CNS are, in part at least, semi-independent. This is perhaps not surprising, as only a small proportion of cerebral axons project into the spinal cord [54]; a histolopathologic case study did report a relationship between cord corticospinal tract axonal loss and acute brain lesions, but this only accounted for 22% of the tissue loss [55].

Several post-mortem studies have characterized the nature of neurodegeneration in the spinal cord [53, 56-65]. There is broad agreement that axonal loss is prominent in the cervical cord, occurs in areas of chronic as well as acute demyelination, and also occurs in normal-appearing tissues. Lesions may involve both GM and WM, and while they do not fully respect anatomical borders, they are most often located in the posterior and lateral columns. In addition, it would appear that small nerve fibers bear the brunt of the damage, where assessed in the cervical cord WM [60]. This suggests that neurodegeneration in the spinal cord is widespread, tract-semi-specific and partly axon-size selective.

Approximately 75%-90% of MS patients have cord lesions visible on T2-weighted MRI, with the cervical spine most likely to be affected [51, 56, 66, 67]. Compared with other clinical phenotypes, primary progressive (PP) MS patients are more likely to develop additional diffuse cord abnormalities on mildly T2-weighted images [51, 68]. Cross-sectional studies have shown no association between the number of spinal cord and brain lesions [51, 66] but a longitudinal study did demonstrate a relationship between the development of new cord and brain lesions in RRMS [69] further emphasizing the complexity of brain- and cord-specific disease mechanisms.

Studies have demonstrated significant decreases in upper cervical cord area (UCCA) in patients with progressive forms of MS in comparison with controls [51, 54, 70-77]. Further, a robust correlation has been found (up to r=–0.7) between UCCA and the EDSS [54, 70, 71, 74]. Cross-sectional studies do not consistently show any significant difference between RRMS phenotypes and controls [52, 70, 73, 76, 78, 79], but longitudinal studies do reveal evidence of a significantly increased rate of atrophy in such patients compared with that of controls [52, 73, 78]. Atrophy rates are low, ranging between 1% and 2% per year

[52, 73, 78]. The difference in atrophy findings between RRMS and progressive subtypes of MS do not appear to simply reflect differences in disease duration, although some previous studies have found this to be relevant [74, 75]. A recent study in RRMS and PPMS both with equivalent and short disease durations noted that only the PP cohort had a significantly smaller UCCA in comparison with controls [80]. This suggests that cord atrophy may be a relatively specific feature of progressive disease. A number of studies reinforce this observation, showing significantly greater degrees of cord atrophy in progressive phenotypes of MS in comparison with their RRMS counterparts [70, 71, 78].

Brain Atrophy Measurement Techniques

Brain atrophy may be estimated from cross-sectional data, by comparing tissue volumes between people with MS and suitably age- and sex-matched control subjects, and with longitudinal data either using the same techniques employed to estimate atrophy from cross-sectional data, or with registration techniques which seek to optimize the match between images derived some time apart, and then determine the differences between them. Such methods can estimate whole-brain, tissue-specific, or regional atrophy. When considering the advantages and disadvantages of each technique, it should be borne in mind that, in the absence of gold-standard intracranial and brain tissue volume measurements in vivo, it is not presently possible to fully assess the accuracy of any methods, and that results should be considered method-specific. Another general caveat is the effects of lesions on volume measurement techniques – the presence of focal abnormalities in brain tissues may systematically affect tissue segmentations such that real disease effects are masked, or artifactual changes introduced [14].

When estimating volumes, it is necessary to define boundaries between regions. MRI sequences may be tuned to maximize tissue differences but, in general, contrast between CSF and brain tissue is greater than that between GM and WM. Contrast at boundaries makes it easier for these to be defined, both manually and automatically, and thus it is more straightforward to measure whole-brain and intracranial volumes compared with GM and WM volumes. This appears to translate into greater reliability for whole-brain compared with tissue-specific volume measures [81].

Given that MRI is not perfect, and that images acquired at different times will have slightly different scaling – i.e., the scanner does not acquire images calibrated for volume – consideration has been given as to how this variability may be reduced. Bone structures in the adult tend not to change rapidly with time, and thus intracranial volume makes a good subject-specific calibration factor. Indeed, this is the basis of most fractional brain tissue measures, such as the brain parenchymal fraction (the ratio of brain tissue to intracranial volume or brain surface contour) [14, 27], or SIENA (structural image evaluation, using normalization, of atrophy) [48]. Such methods appear to differ marginally in their reliability, but more noticeably in terms of accuracy [82]. While these techniques may be applied to both cross-sectional and serial data, methods have been developed that look directly for changes in tissue volumes. These rely on constrained image registration, followed by an assessment of the differences between the images [5, 48]. Within the GM, methods have also been devised to regionally localize consistent atrophy [12, 17, 18, 83]. Registration-based methods appear more reliable than segmentation-subtraction techniques in detecting small degrees of atrophy on serial images [84].

Cord Atrophy Measurement Techniques

Imaging of the cord presents particular challenges. The cord is a small and relatively mobile structure, and image quality may be degraded by ghosting artifacts from the heart and large vessels, and by truncation artifacts [85]. The spinal cord's small diameter and its position surrounded by bone also lower the signal-to-noise ratio. MR developments such as cardiac gating, spatial presaturation slabs, and fast imaging sequences which still provide good resolution, such as three-dimensional fast spoiled gradient echo recall (3D-FSPGR) have been employed to improve the accuracy of measures of cord atrophy. However, problems such as possible partial volume error at the cord/CSF boundary are a potential source of error and may lead to overestimation of cord area values [70]. Methods are being developed to minimize this, but they are still at relatively early stages of development [86, 87]. However, the Losseff technique (described below) has shown itself capable of detecting low rates of atrophy longitudinally in MS [52].

The earliest estimations of cross-sectional cord area required manual outlining on axial images, usually at the level of C5 generated by a two-dimensional gradient echo sequence [66, 71]. The large degree of operator input impacted negatively on reproducibility. With improved volume sequences (3D-FSPGR), a semi-automated measurement technique was developed using images reformatted with slices perpendicular to the cord at the level of C2/3 [70]. The technique is based on the differing signal intensity signatures exhibited by cord tissue and CSF. A contour is drawn at signal intensity halfway between that of cord and CSF and this corresponds to the boundary of the cord. Area is estimated as an average over five contiguous 3-mm axial slices. The level C2/3 is used because the CSF space tends to be wide here; it is generally easier to position the patient to maximize the CSF space in this location and also it is an uncommon site for disc protrusion. This technique offers excellent reproducibility (coefficient of variation 0.79% compared with 6% as seen in manual methods).

Other cord area measurement techniques that have been employed include the Sobell technique, which uses a similar 3D sequence but detects partial volume boundary automatically between cord and CSF [54], and recent modifications to this which have attempted to minimize potential partial volume inaccuracy [87].

Cord volume has also been measured using a variation of the Sobell technique [26, 74] and using a semi-automatic method based on optimization of a B-spline active surface model of the cord [86]. At present these volume measures do not appear to confer any significant advantage over the less computationally intensive area techniques.

Studies have looked into differing clinical phenotypes of MS and have also attempted to delineate any relationships between the measured data and correlates of clinical disability (see below). One of the primary objectives of these studies is to determine whether atrophy as a quantitative measure is a suitable surrogate marker for disease progression in MS. The attractions of this are numerous, not least the potential to improve the reproducibility and sensitivity of outcome measures in clinical treatment trials [88]. However, in order for MR measures of atrophy to be considered for such a role, the measure should be functionally relevant and dynamically reflect clinical endpoints. Ongoing quantitative studies measuring atrophy aim, by improving the accuracy of findings, to satisfy these criteria. As brain and spinal cord tissue loss may be the result of a differing combination of pathological processes, as described earlier, it is logical that the relevance of these measures may differ depending on the stage of MS that is being quantified.

Study design is also important. Additional factors are known to be potentially associated to cord area values and therefore must be included in any statistical modeling when cohorts are compared. These include estimates of total intracranial volume, gender, and possibly patient height [70, 78, 79, 89].

Previous Applications of Atrophy Measures in MS Treatment Trials

Increasingly, brain atrophy measures are being employed as secondary endpoints in MS treatment trials. The correlations between this measure and clinical disability and the attraction of a quantitative measure in addition to clinical scales, as described earlier in this chapter, are key reasons for this. However, studies looking for an association between brain and cord tissue loss and their modification by disease-modifying therapies in MS, such as interferon β (IFNβ), have been unclear.

The picture regarding brain atrophy is somewhat mixed. Studies looking at the weekly administration of IFNβ-1a in RRMS patients have shown a possible slowing of rate of tissue loss in two longitudinal studies ranging between 2 and 3 years [90, 91] and significant benefit in a study investigating once weekly IFNβ-1a administration for 2 years [92]. Conversely, a study using IFNβ-1a on PPMS patients showed no benefit regarding measures of ventricular volume [93], whilst studies employing the more commonly used three times weekly administration of IFNβ-1a in RRMS patients have shown varied results; no significant benefit in two studies [94, 95] and a possible benefit in another [11]. IFNβ-1b studies in RRMS and SPMS have both shown no benefit regarding atrophy progression [96, 97]. Glatimer acetate studies in RRMS have shown differing results in four trials; two showing possible benefits and two not [75, 98-100].

Effects of disease-modifying therapy on spinal cord area are possibly even less clear. At present there is a paucity of such studies, and those undertaken have been hampered by small cohort numbers. A subset of a three times weekly IFNβ-1a study in 38 patients with either RRMS or SPMS did not show any statistical benefit of the drug on UCCA [78], whilst a study using IFNβ-1a on 50 PPMS patients again showed no benefit on the rate of tissue loss [93]. Of possible interest, a potential benefit has been shown in a pilot study on 16 patients with PPMS using riluzole on the rate of cord atrophy [101]. This requires further confirmation.

These varied results illustrate the uncertainty of the effect of disease-modifying therapy on tissue loss. The complexity of pathological mechanisms responsible for atrophy may in part be responsible for this. In addition, the effect of medication may, at least in the early stages of therapy, make atrophy more noticeable because of the potential resolution of inflammation and edema, and consequent reduction in tissue swelling associated with this. Further, trials measuring spinal cord atrophy are few and patient numbers small. This emphasizes the need for larger trials of longer duration in order to determine the effect of therapy on tissue loss.

As sample sizes are clarified for the numbers needed to treat to show a reduction in the rate of brain atrophy, proof-of-concept trials are anticipated that use brain atrophy as the primary outcome measure. Such a trial of neuroprotection in secondary progressive MS has recently got under way [102].

Conclusions

The mechanisms involved in atrophy are complex. They partly reflect axonal loss, whether it be through a secondary process such as anterograde or retrograde degeneration, primary neural pathology, or a combination of both. The process is almost certainly multifactorial, and is likely to reflect clinical variability in MS. Both brain and spinal cord measures of atrophy have proved to be clinically relevant, and there is a suggestion that measures of tissue loss in the brain and cord may be complementary, given that regional atrophy may have different clinical consequences. Brain atrophy measures appear to be more sensitive than equivalent cord values to significant tissue loss earlier in the course of the disease [16, 24, 38, 52, 79].

In general, correlations with EDSS are generally stronger with UCCA than with brain atrophy. This may reflect the nature of the EDSS as a measure of disability weighted towards lower limb function, and thus perhaps more directly influenced by cord pathology, while regional brain atrophy may in part result in deficits in cognition [103] and fatigue [104]. Newer treatment approaches will hopefully seek to address these distressing symptoms, and will almost certainly require quantitative outcome measures to evaluate their efficacy.

Both brain and cord imaging techniques are increasingly readily available, and accuracy of analysis is improving. With more potential therapies becoming available for MS in the future, the need for quantitative in vivo outcome measures, such as atrophy, has never been greater.

Acknowledgements. The NMR Unit at The Institute of Neurology is supported by the Multiple Sclerosis Society of Great Britain and Northern Ireland. No competing interests are declared.

References

1. Peterson JW, Bö L, Monk S et al (2001) Transected neuritis, apoptotic neurons, and reduced inflammation in cortical multiple sclerosis lesions. Ann Neurol 50:389-400
2. Bo L, Vedeler CA, Nyland H et al (2003) Intracortical multiple sclerosis lesions are not associated with increased lymphocyte infiltration. Mult Scler 9:323-331
3. Brink BP, Veerhuis R, Breij EC et al (2005) The pathology of multiple sclerosis is location-dependent: no significant complement activation is detected in purely cortical lesions. J Neuropathol Exp Neurol 64:147-155
4. Brex PA, Jenkins R, Fox NC et al (2000) Detection of ventricular enlargement in patients at the earliest clinical stage of MS. Neurology 54:1689-1691
5. Dalton CM, Brex PA, Jenkins R et al (2002) Progressive ventricular enlargement in patients with clinically isolated syndromes is associated with the early development of multiple sclerosis. J Neurol Neurosurg Psychiat 73:141-147
6. Dalton CM, Chard DT, Davies GR et al (2004) Early development of multiple sclerosis is associated with progressive grey matter atrophy in patients presenting with clinically isolated syndromes. Brain 127:1101-1107
7. Ge Y, Grossman RI, Udupa JK et al (2000) Brain atrophy in relapsing-remitting multiple sclerosis and secondary progressive multiple sclerosis: longitudinal quantitative analysis. Radiology 214:665-670
8. Bakshi R, Benedict RH, Bermel RA, Jacobs L (2001) Regional brain atrophy is associated with physical disability in multiple sclerosis: semiquantitative magnetic resonance imaging and relationship to clinical findings. J Neuroimaging 11:129-136
9. Lin X, Blumhardt LD (2001) Inflammation and atrophy in multiple sclerosis: MRI associations with disease course. J Neurol Sci 189:99-104
10. Kalkers NF, Ameziane N, Bot JC et al (2002) Longitudinal brain volume measurement in multiple sclerosis: rate of brain atrophy is independent of the disease subtype. Arch Neurol 59:1572-1576
11. Turner B, Lin X, Calmon G et al (2003) Cerebral atrophy and disability in relapsing-remitting and secondary progressive multiple sclerosis over four years. Mult Scler 9:21-27
12. Pagani E, Rocca MA, Gallo A et al (2005) Regional brain atrophy evolves differently in patients with multiple sclerosis according to clinical phenotype. Am J Neuroradiol 26:341-346
13. Tedeschi G, Lavorgna L, Russo P et al (2005) Brain atrophy and lesion load in a large population of patients with multiple sclerosis. Neurology 65:280-285
14. Chard DT, Griffin CM, Parker GJM et al (2002) Brain atrophy in clinically early relapsing-remitting multiple sclerosis. Brain 125:327-337
15. Chard DT, Griffin CM, Rashid W et al (2004) Progressive grey matter atrophy in clinically early relapsing-remitting multiple sclerosis. Mult Scler 10:387-391
16. Tiberio M, Chard DT, Altmann DR et al (2005) Grey and white matter volume changes in early RRMS: a two-year longitudinal study. Neurology 65:1001-1007
17. Sailer M, Fischl B, Salat D et al (2003) Focal thinning of the cerebral cortex in multiple sclerosis. Brain 126:1734-1744
18. Audoin B, Davies GR, Fishniku L et al (2006) Localization of grey matter atrophy in early RRMS: a longitudinal study. J Neurol 253:1495-1501

19. Carone DA, Benedict RH, Dwyer MG et al (2006) Semi-automatic brain region extraction (SABRE) reveals superior cortical and deep gray matter atrophy in MS. NeuroImage 29:505-514
20. Prinster A, Quarantelli M, Orefice G et al (2006) Grey matter loss in relapsing-remitting multiple sclerosis: a voxel-based morphometry study. NeuroImage 29:859-867
21. Cifelli A, Arridge M, Jezzard P et al (2002) Thalamic neurodegeneration in multiple sclerosis. Ann Neurol 52:650-653
22. Redmond IT, Barbosa S, Blumhardt LD, Roberts N (2000) Short-term ventricular volume changes on serial MRI in multiple sclerosis. Acta Neurol Scand 102:99-105
23. Fox NC, Jenkins R, Leary SM et al (2000) Progressive cerebral atrophy in MS: a serial study using registered, volumetric MRI. Neurology 54:807-812
24. Dalton CM, Miszkiel KA, O'Connor PW et al (2006) Ventricular enlargement in MS: one-year change at various stages of disease. Neurology 66:693-698
25. Benedict RH, Bruce JM, Dwyer MG et al (2006) Neocortical atrophy, third ventricular width, and cognitive dysfunction in multiple sclerosis. Arch Neurol 63:1301-1306
26. Liu C, Edwards S, Gong Q et al (1999) Three dimensional MRI estimates of brain and spinal cord atrophy in multiple sclerosis. J Neurol Neurosurg Psychiat 66:323-330
27. Rudick RA, Fisher E, Lee JC et al (2000) Brain atrophy in relapsing multiple sclerosis: relationship to relapses, EDSS and treatment with interferon beta-1a. Mult Scler 6:365-372
28. Zivadinov R, Locatelli L, Stival B et al (2003) Normalized regional brain atrophy measurements in multiple sclerosis. Neuroradiology 45:793-798
29. Quarantelli M, Ciarmiello A, Morra VB et al (2003) Brain tissue volume changes in relapsing-remitting multiple sclerosis: correlation with lesion load. NeuroImage 18:360-366
30. Sastre-Garriga J, Ingle GT, Chard DT et al (2004) Grey and white matter atrophy in early clinical stages of primary progressive multiple sclerosis. NeuroImage 22:353-359
31. Filippi M, Rovaris M, Iannucci G et al (2000) Whole brain volume changes in patients with progressive MS treated with cladribine. Neurology 55:1714-1718
32. Paolillo A, Pozzilli C, Gasperini C et al (2000) Brain atrophy in relapsing-remitting multiple sclerosis: relationship with 'black holes', disease duration and clinical disability. J Neurol Sci 174:85-91
33. Kalkers NF, Vrenken H, Uitdehaag BM et al (2002) Brain atrophy in multiple sclerosis: impact of lesions and of damage of whole brain tissue. Mult Scler 8:410-414
34. De Stefano N, Iannucci G, Sormani MP et al (2002) MR correlates of cerebral atrophy in patients with multiple sclerosis. J Neurol 249:1072-1077
35. Kassubek J, Tumani H, Ecker D et al (2003) Age-related brain parenchymal fraction is significantly decreased in young multiple sclerosis patients: quantitative MRI study. Neuroreport 14:427-430
36. Simon JH (1999) From enhancing lesions to brain atrophy in relapsing MS. J Neuroimmunol 98:7-15
37. Fisher E, Rudick RA, Simon JH et al (2002) Eight-year follow-up study of brain atrophy in patients with MS. Neurology 59:1412-1420
38. Chard DT, Brex PA, Ciccarelli O et al (2003) The longitudinal relation between brain lesion load and atrophy in multiple sclerosis: a 14 year follow up study. J Neurol Neurosurg Psychiat 74:1551-1554
39. Dastidar P, Heinonen T, Lehtimaki T et al (1999) Volumes of brain atrophy and plaques correlated with neurological disability in secondary progressive multiple sclerosis. J Neurol Sci 165:36-42
40. Simon JH, Jacobs LD, Campion MK et al (1999) A longitudinal study of brain atrophy in relapsing multiple sclerosis. The Multiple Sclerosis Collaborative Research Group (MSCRG). Neurology 53:139-148

41. Fisher E, Rudick RA, Cutter G et al (2000) Relationship between brain atrophy and disability: an 8-year follow-up study of multiple sclerosis patients. Mult Scler 6:373-377
42. Zivadonov R, Sepcic J, Nasuelli D et al (2001) A longitudinal study of brain atrophy and cognitive disturbances in the early phase of relapsing-remitting multiple sclerosis. J Neurol Neurosurg Psychiatry 70:773-780
43. Paolillo A, Pozzilli C, Giugni E et al (2002) A 6-year clinical and MRI follow-up study of patients with relapsing-remitting multiple sclerosis treated with interferon-beta. Eur J Neurol 9:645-655
44. Bermel RA, Sharma J, Tjoa CW et al (2003) A semiautomated measure of whole-brain atrophy in multiple sclerosis. J Neurol Sci 208:57-65
45. Locatelli L, Zivadinov R, Grop A, Zorzon M (2004) Frontal parenchymal atrophy measures in multiple sclerosis. Mult Scler 10:562-568
46. Zivadinov R, Bakshi R (2004) Central nervous system atrophy and clinical status in multiple sclerosis. J Neuroimaging 14:27S-35S
47. Sanfilipo MP, Benedict RH, Sharma J et al (2005) The relationship between whole brain volume and disability in multiple sclerosis: a comparison of normalized gray vs. white matter with misclassification correction. NeuroImage 26:1068-1077
48. Sastre-Garriga J, Ingle GT, Chard DT et al (2005) Grey and white matter volume changes in early primary progressive multiple sclerosis: a longitudinal study. Brain 128:1454-1460
49. Sanfilipo MP, Benedict RH, Weinstock-Guttman B, Bakshi R (2006) Gray and white matter brain atrophy and neuropsychological impairment in multiple sclerosis. Neurology 66:685-692
50. Kurtzke JF (1983) Rating neurologic impairment in multiple sclerosis: an expanded disability status scale (EDSS). Neurology 33:1444-1452
51. Nijeholt GJ, van Walderveen MAA, Castelijns JA et al (1998) Brain and spinal cord abnormalities in multiple sclerosis: correlations between MRI parameters, clinical subtypes and symptoms. Brain 121:687-697
52. Rashid W, Davies GR, Chard DT et al (2006) Increasing cord atrophy in early relapsing-remitting multiple sclerosis: a 3 year study. J Neurol Neurosurg Psychiat 77:51-55
53. DeLuca GC, Williams K, Evangelou N et al (2006) The contribution of demyelination to axonal loss in multiple sclerosis. Brain 129:1507-1516
54. Lin X, Tench CR, Evangelou N et al (2004) Measurement of spinal cord atrophy in multiple sclerosis. J Neuroimaging 14:20S-26S
55. Bjartmar C, Kinkel RP, Kidd G et al (2001) Axonal loss in normal-appearing white matter in a patient with acute MS. Neurology 57:1248-1252
56. Oppenheimer DR (1978) The cervical cord in multiple sclerosis. Neuropathol Appl Neurobiol 4:151-162
57. Ganter P, Prince C, Esiri MM (1999) Spinal cord axonal loss in multiple sclerosis: a post-mortem study. Neuropathol Appl Neurobiol 25:459-467
58. McGavern DB, Murray PD, Rivera-Quiñones C et al (2000) Axonal loss results in spinal cord atrophy, electrophysiological abnormalities and neurological deficits following demyelination in a chronic inflammatory model of multiple sclerosis. Brain 123:519-531
59. Bjartmar C, Kidd G, Mörk S et al (2000) Neurological disability correlates with spinal cord axonal loss and reduced N-acetyl aspartate in chronic multiple sclerosis patients. Ann Neurol 48:893-901
60. Lovas G, Szilagyi N, Majtenyi K et al (2000) Axonal changes in chronic demyelinated cervical spinal cord plaques. Brain 123:308-317
61. Nijeholt GJ, Bergers E, Kamphorst W et al (2001) Post-mortem high resolution MRI of the spinal cord in multiple sclerosis: a correlative study with conventional MRI, histopathology and clinical phenotype. Brain 124:154-166

62. Bergers E, Bot JC, De Groot CJ et al (2002) Axonal damage in the spinal cord of MS patients occurs largely independent of T2 MRI lesions. Neurology 59:1766-1771
63. DeLuca GC, Ebers GC, Esiri MM (2004) Axonal loss in multiple sclerosis: a pathological survey of the corticospinal and sensory tracts. Brain 127:1009-1018
64. Gilmore CP, DeLuca GC, Bö L et al (2005) Spinal cord atrophy in multiple sclerosis caused by white matter volume loss. Arch Neurol 62:1859-1862
65. Gilmore CP, Bö L, Owens T et al (2006) Spinal cord gray matter demyelination in multiple sclerosis: a novel pattern of residual plaque morphology. Brain Pathol 16:202-208
66. Kidd D, Thorpe JW, Thompson AJ et al (1993) Spinal cord MRI using multi-array coils and fast spin echo. II. Findings in multiple sclerosis. Neurology 43:2632-2637
67. Bot JC, Barkhof F, Polman CH et al (2004) Spinal cord abnormalities in recently diagnosed MS patients: added value of spinal MRI examination. Neurology 62:226-233
68. Lycklama à Nijeholt GJ, Barkhof F, Scheltens P et al (1997) MR of the spinal cord in multiple sclerosis: relation to clinical subtype and disability. Am J Neuroradiol 18:1041-1048
69. Thorpe JW, Kidd D, Moseley IF et al (1996) Serial gadolinium-enhanced MRI of the brain and spinal cord in early relapsing-remitting multiple sclerosis. Neurology 46:373-378
70. Losseff NA, Webb SL, O'Riordan JI et al (1996) Spinal cord atrophy and disability in multiple sclerosis: a new reproducible and sensitive MRI method with potential to monitor disease progression. Brain 119:701-708
71. Filippi M, Campi A, Colombo B et al (1996) A spinal cord MRI study of benign and secondary progressive multiple sclerosis. J Neurol 243:502-505
72. Filippi M, Colombo B, Rovaris M et al (1997) A longitudinal magnetic resonance imaging study of the cervical cord in multiple sclerosis. J Neuroimaging 7:78-80
73. Stevenson VL, Leary SM, Losseff NA et al (1998) Spinal cord atrophy and disability in MS: a longitudinal study. Neurology 51:234-238
74. Edwards SGM, Gong QY, Lui C et al (1999) Infratentorial atrophy on magnetic resonance imaging and disability in multiple sclerosis. Brain 122:291-301
75. Rovaris M, Bozzali M, Santuccio G et al (2001) In vivo assessment of the brain and cervical cord pathology with primary progressive multiple sclerosis. Brain 124:2540-2549
76. Vaithianathar L, Tench CR, Morgan PS, Constantinescu CS (2003) Magnetic resonance imaging of the cervical spinal cord in multiple sclerosis: a quantitative T1 relaxation time mapping approach. J Neurol 250:307-315
77. Lin X, Tench CR, Turner B et al (2003) Spinal cord atrophy and disability in multiple sclerosis over four years: application of a reproducible automated technique in monitoring disease progression in a cohort of the interferon β-1a (Rebif) treatment trial. J Neurol Neurosurg Psychiatry 74:1090-1094
78. Lin X, Blumhardt LD, Constantinescu CS (2003) The relationship of brain and cervical cord volume to disability in clinical subtypes of multiple sclerosis: a three-dimensional MRI study. Acta Neurol Scand 108:401-406
79. Rashid W, Davies GR, Chard DT et al (2006) Upper cervical cord area in early relapsing-remitting multiple sclerosis: cross-sectional study of factors influencing cord size. J Magn Reson Imaging 23:473-476
80. Beniek M, Altmann DR, Davies GR et al (2006) Cord atrophy separates early primary progressive and relapsing multiple sclerosis. J Neurol Neurosurg Psychiat 77:1036-1039
81. Chard DT, Parker GJ, Griffin CM et al (2002) The reproducibility and sensitivity of brain tissue volume measurements derived from an SPM-based segmentation methodology. J Magn Reson Imaging 15:259-267

82. Zivadinov R, Grop A, Sharma J et al (2005) Reproducibility and accuracy of quantitative magnetic resonance imaging techniques of whole-brain atrophy measurement in multiple sclerosis. J Neuroimaging 15:27-36
83. Charil A, Dagher A, Lerch JP et al (2007) Focal cortical atrophy in multiple sclerosis: relation to lesion load and disability. NeuroImage 34:509-517
84. Anderson VM, Fernando KT, Davies GR et al (2007) Cerebral atrophy measurement in clinically isolated syndromes and relapsing remitting multiple sclerosis: a comparison of registration-based methods. J Neuroimaging 17:61-68
85. Lycklama G, Thompson A, Filippi M et al (2003) Spinal-cord MRI in multiple sclerosis. Lancet Neurol 2:555-562
86. Coulon O, Hickman SJ, Parker GJ et al (2002) Quantification of spinal cord atrophy from magnetic resonance images via a B-spline active surface model. Magn Reson Med 47:1176-1185
87. Tench CR, Morgan PS, Constantinescu CS (2005) Measurement of cervical spinal cord cross-sectional area by MRI using edge detection and partial volume correction. J Magn Reson Imaging 21:197-203
88. Filippi M, Horsfield MA, Ader HJ et al (1998) Guidelines for using quantitative measures of brain magnetic resonance imaging abnormalities in monitoring the treatment of multiple sclerosis. Ann Neurol 43:499-506
89. Kameyama T, Hashizume Y, Ando T, Takahashi A (1994) Morphometry of the normal cadaveric spinal cord. Spine 19:2077-2081
90. Rudick RA, Fischer E, Lee J-C et al (1999) Use of brain parenchymal fraction to measure whole brain atrophy in relapsing-remitting MS. Neurology 53:1698-1704
91. Hardmeier M, Freitag P, Wagenpfeil et al (2002) Short and long term brain volume changes after initiation of treatment with rIFN-beta-1A in multiple sclerosis (MS) J Neurol 249(Suppl 1):20 (abstract)
92. Filippi M, Rovaris M, Inglese M et al and the ETOMS Study Group (2004) Interferon beta-1a for brain tissue loss in patients at presentation with syndromes suggestive of multiple sclerosis: a randomised, double-blind, placebo-controlled trial. Lancet 364:1489-1496
93. Leary SM, Miller DH, Stevenson VL et al (2003) Interferon β-1a in primary progressive MS: an exploratory, randomized, controlled trial. Neurology 60:44-51
94. Jones CK, Riddehough A, Li DKB et al (2001) MRI cerebral atrophy in relapsing-remitting MS: results from the PRISMS trial. Neurology 56(Suppl 3):A379
95. Gasperini C, Paolillo A, Giugni E et al (2002) MRI brain volume changes in relapsing-remitting multiple sclerosis patients treated with interferon beta-1a. Mult Scler 8:119-123
96. Molyneux PD, Kappos L, Polman C et al (2000) The effect of interferon beta-1b treatment on MRI measures of cerebral atrophy in secondary progressive multiple sclerosis. Brain 123:2256-2263
97. Frank JA, Richert N, Bash C et al (2004) Interferon-β-1b slows progression of atrophy in RRMS: three-year follow-up in NAb^- and NAb^+ patients. Neurology 62:719-725
98. Ge Y, Grossman RI, Udupa JK et al (2000) Glatiramer acetate (Copaxone) treatment in relapsing-remitting MS: quantitative MR assessment. Neurology 54:813-817
99. Wolinsky JS, Narayana PA, Johnson KP (2001) United States open-label glatiramer acetate extension trial for relapsing multiple sclerosis: MRI and clinical correlates. Copolymer 1 Multiple Sclerosis Study Group, MRI Analysis Center. Mult Scler 7:33-41
100. Sormani MP, Rovaris M, Valsasina P et al (2004) Measurement error of two different techniques for brain atrophy assessment in multiple sclerosis. Neurology 62:1432-1434
101. Kalkers NF, Barkhof F, Bergers E et al (2002) The effect of the neuroprotective agent riluzole on MRI parameters in primary progressive multiple sclerosis: a pilot study. Mult Scler 8:532-533

102. Furby J, Hayton T, Smith KJ et al (2006) A randomised controlled trial of neuroprotection with lamotrigine in secondary progressive multiple sclerosis. Mult Scler 12 (S228):P794
103. Morgen K, Sammer G, Courtney SM et al (2006) Evidence for direct association between cortical atrophy and cognitive impairment in relapsing-remitting MS. Neuroimage 30:891-898
104. Marrie RA, Fisher E, Miller DM et al (2005) Association of fatigue and brain atrophy in multiple sclerosis. J Neurol Sci 228:161-166

Chapter 4

T1 Black Holes and Gray Matter Damage

M. Neema, V.S.R. Dandamudi, A. Arora, J. Stankiewicz, R. Bakshi

Introduction

Magnetic resonance imaging (MRI) has become important in the early diagnosis and monitoring of various neurologic disorders including multiple sclerosis (MS). MRI has emerged as a key supportive therapeutic outcome measure in MS-related clinical trials. The limitations of conventional MRI surrogates have driven researchers to develop better biomarkers, including those capturing destructive aspects of the disease. In this chapter, we discuss the most recent data highlighting the role of hypointense lesions on T1-weighted images (black holes; BH) and gray matter (GM) damage in the MRI assessment of MS. We focus on the most relevant pathologic, MRI, and clinical correlation studies addressing BH and GM injury.

T1 Black Holes

MRI Findings

A subset of MS plaques seen on conventional T2-weighted and fluid-attenuated inversion-recovery (FLAIR) images in the brain may appear hypointense on corresponding T1-weighted images in comparison with the surrounding normal-appearing white matter (NAWM) (Fig. 1). These T1-hypointense foci are commonly referred to as "black holes" (BHs). T2 or FLAIR hyperintense areas may correspond to all or only a portion of a BH. The degree of T1-hypointensity is dependent on the scan parameters in the MRI pulse sequence, which may account for differences in the results reported by different investigators. BHs initially start as gadolinium (Gd) enhancing lesions, and evolve into two categories of lesions: transient (reversible) or chronic (persistent), as discussed below. BHs are rarely seen in the posterior fossa and spinal cord. Longitudinal studies have shown that the amount and duration of Gd-enhancement at baseline is predictive of the development of persistent BHs [1]. However, due to the variable natural history of evolution of BHs and Gd-enhancement, the predictive power is limited.

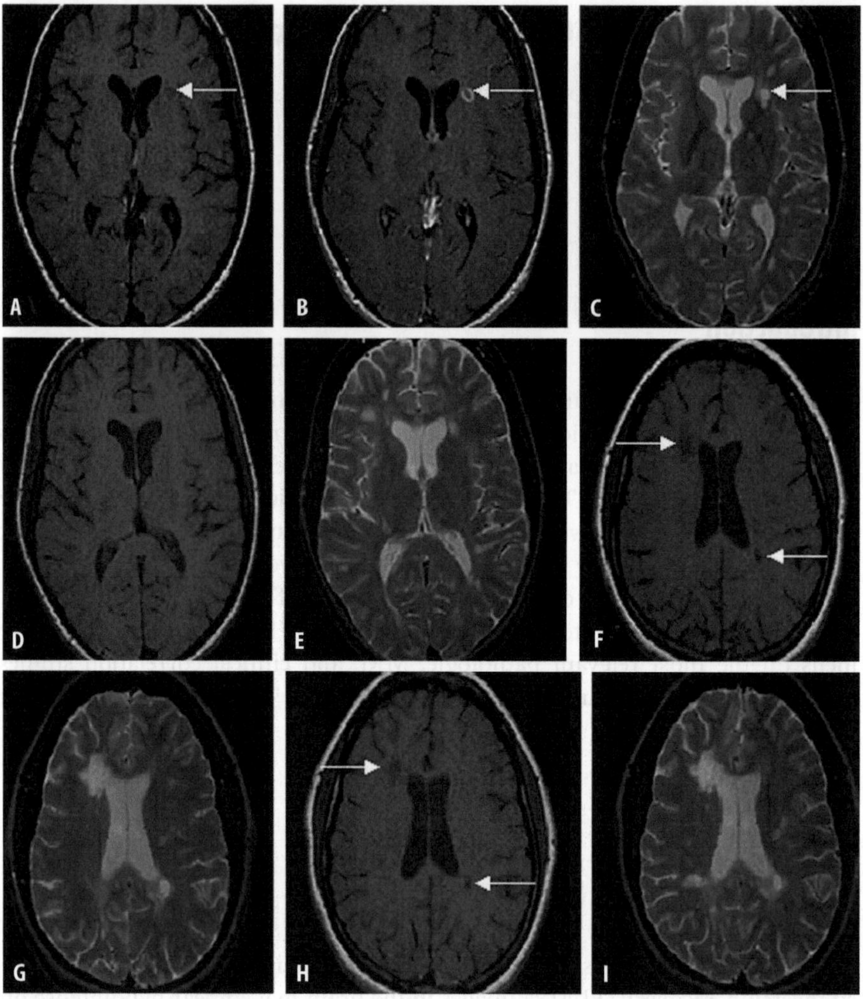

Fig. 1. Black holes (BHs) on T1-weighted images: resolving (*top panel*) and chronic (*lower panel*). **A, D, F,** and **H** are conventional spin-echo (CSE) T1-weighted images. **B** is a post-contrast (gadolinium) T1-weighted image. **C, E, G,** and *I* are CSE T2-weighted images. **A–E** are from a patient with relapsing-remitting MS. The baseline images (**A–C**) show a newly formed BH (*arrow*). Eight months later (**D, E**), the BH has resolved on the T1-weighted image. *F-I* are from a patient with relapsing-remitting MS. The baseline images (**F, G**) show two BHs (*arrows*). Nine months later (**H, I**), the BHs are persistent on the T1-weighted image (*arrows*)

Neuropathology

BHs most probably reflect variable combinations of inflammation, edema, demyelination, early remyelination, axonal transection, and glial (microglial/astrocyte) activation [2]. These pathological correlates are dependent on the extent of hypointensity and the age of the lesion. BHs that show profound hypointensity on T1-weighted post-mortem images correlate pathologically with severe demyelination and axonal loss [3]. During the natural disease course, about half of the newly formed BHs disappear within 6-12 months and revert back to isointensity as a result of resolution of edema and remyelination (Fig. 1). BHs that persist for more than 12 months reflect severe demyelination and axonal loss [5] (Fig. 1). The tendency for newly formed lesions to develop into chronic persistent BH has been linked to the APOE-e4 allele [5]. Findings with advanced MRI techniques, such as decreased magnetization transfer ratio (MTR), can potentially predict the conversion from acute to persistent BHs [6].

Clinical Correlation

Various groups have shown moderate to strong clinical correlations in both cross-sectional and longitudinal studies between BHs and neurologic disability as measured by expanded disease severity status (EDSS) in patients with relapsing/remitting (RR) MS and secondary progressive (SP) MS. Global BH volume shows a somewhat better correlation with clinical status than does conventional T2 lesion load [7-9] (Table 1). The interpretation of whole brain BH volume in therapeutic studies is complicated by the potential for resolution of individual BHs. Longitudinal analysis of the evolution of each newly formed BH can overcome this limitation [10, 11].

Role in Therapeutic Monitoring

Investigators have assessed change in global BH lesion volume and the evolution of individual BHs in order to explore drug efficacy in clinical trials. In RRMS or SPMS patient cohorts from randomized, placebo-controlled, prospective studies, immunomodulatory agents such as interferon beta (IFNB) [12, 13] and glatiramer acetate (GA) [14] have shown modest effects on attenuating global BH

Table 1. Correlations between MRI measures (atrophy and lesion load) and physical disability in patients with multiple sclerosis

Variable	Expanded Disability Status Scale	Timed 25-foot walk
n	40	35
Global parenchymal volume	– 0.34*	– 0.45**
Global gray matter volume	– 0.46**	– 0.52**
Global white matter volume	– 0.06	– 0.20
Third ventricle width	0.20	0.43*
Bicaudate ratio	0.32*	0.32
Global T1 black hole lesion volume	0.41**	0.46**
Global FLAIR lesion volume	0.34*	0.38*

Partial correlations are shown expressing relationships after adjusting for intracranial volume and age. FLAIR = fluid-attenuated inversion-recovery; *n* = number of patients.
*$P < 0.05$; **$P < 0.01$ (adapted from [8])

volume increases. In addition, trials with GA [10] and natalizumab [11] have shown that Gd-enhancing lesions, which developed despite therapy, were less likely to develop as chronic BHs in patients with RRMS.

Summary and Future Directions

Research investigating the substrate and mechanisms underlying the development and evolution of BHs has made progress recently. Assaying for BHs in white matter (WM) improves correlation with neurological impairment when compared to T2-weighted conventional MRI lesion measures. However, these interpretations are complicated by the variable evolution of newly formed BHs. The conversion from newly formed to chronic BHs serves as a useful surrogate for disease progression and the effectiveness of treatment modalities. Additional studies uncovering the mechanisms involved in BH evolution should provide a better understanding of the destructive aspects of MS.

Gray Matter Involvement

Neuropathology

Although MS has been considered a WM disease, many studies have reported involvement of gray matter (GM), including the cerebral cortex and deep cerebral nuclei, beginning early in the disease course. Post-mortem brain histology studies show cortical lesions that tend to generally, but not exclusively, follow venous structures. Cortical lesions have been subdivided into morphologic subtypes based on histologic characteristics and location [15, 16]. GM lesions show demyelination, microglial activation, and neuronal apoptosis [16, 17]. There are proposed differences in the pathology of WM and GM lesions. GM lesions are thought to be less inflammatory than WM lesions, with a dominant effector cell population of ramified microglia in GM [16, 18]. A study from another group [17], using biopsied tissue in newly diagnosed patients with MS shows contrasting results, including the presence of immune cell infiltration in cortical lesions. A component of GM involvement also appears to include diffuse pathology affecting non-lesional areas (affecting non-lesional WM as well) that is dissociated from overt inflammatory demyelinating foci in the WM [19]. Finally, another manifestation of GM involvement is neuronal loss occurring secondary to axonal injury in WM, leading to Wallerian degeneration. Thus, it is likely that both primary and secondary mechanisms of injury affect the GM in patients with MS.

Imaging Findings

MRI studies have shown GM involvement in patients with MS, including lesions, atrophy, and T2 hypointensity. Cortical lesions are poorly seen on conventional MRI sequences but can be detected with more sensitivity by FLAIR (Fig. 2) and double inversion recovery [20, 21].

MRI segmentation techniques provide evidence for both global and regional GM atrophy in patients with MS [22]. In some - but not all - studies, global GM atrophy is selective (i.e., out of proportion) when compared to WM or global cerebral atrophy [8, 23]. The disproportionate loss of GM versus WM volume may relate in part to episodic inflammation and edema that increases WM volume and masks ongoing atrophy in WM [23]. A significant loss of GM volume is seen early in the disease course, such as in patients with a first attack of demyelination [23] and in mildly disabled patients with established RRMS [24]. Such global GM volume loss can be detected in as little as a 9-month observation period [24]. Using regional atrophy measurement techniques, volume loss of both cortical and subcortical GM areas has been documented in patients with MS as compared to normal controls, such as in the frontotemporal cortex, precuneus, anterior cingulated gyrus, caudate, and thalamus [25-27].

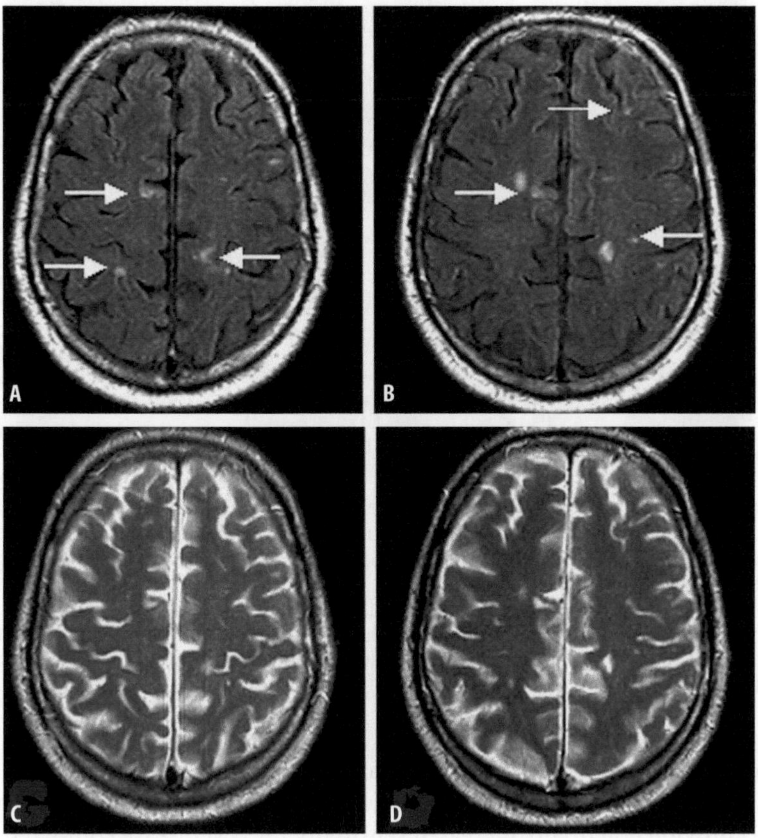

Fig. 2. Fluid-attenuated inversion-recovery (FLAIR) images (**A**, **B**) and fast-spin echo T2-weighted images (**C**, **D**) of a 53-year-old man with relapsing/remitting MS. Note cortical and juxtacortical lesions (*arrows*) are clearly visible on FLAIR images but poorly seen on T2-weighted images

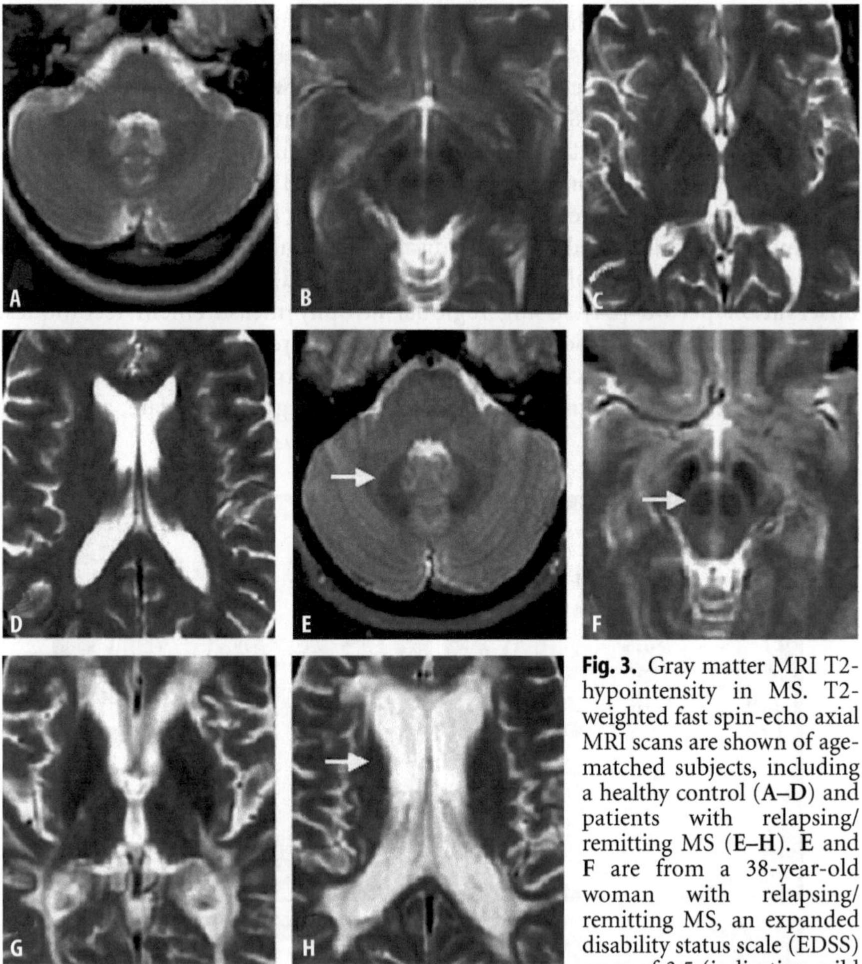

Fig. 3. Gray matter MRI T2-hypointensity in MS. T2-weighted fast spin-echo axial MRI scans are shown of age-matched subjects, including a healthy control (**A–D**) and patients with relapsing/remitting MS (**E–H**). **E** and **F** are from a 38-year-old woman with relapsing/remitting MS, an expanded disability status scale (EDSS) score of 3.5 (indicating mild to moderate physical disability), and a disease duration of 10 years. Note the hypointensity of the dentate nuclei (**E**, *arrow*) and red nuclei (**F**, *arrow*) in comparison with the normal control (**A**, **B**). **G** and **H** are from a 44-year-old man with relapsing/remitting MS, moderate disability (EDSS score 5.5), and a disease duration of 9 years. In **G**, note the hypointensity of the thalamus, putamen, and globus pallidus compared with the normal control (**C**). In **H**, note the hypointensity of the caudate nuclei (*arrow*) compared with the normal control (**D**). The patient also has evidence of diffuse brain atrophy including ventricular and sulcal enlargement compared with the normal subject

Advanced MRI techniques such as magnetic resonance spectroscopy (MRS), magnetization transfer ratio (MTR), and diffusion tensor imaging (DTI) provide potentially increased sensitivity for GM disease compared with conventional MRI methods in patients with MS. GM shows reduced N-acetylaspartate (NAA) [28] and decreased MTR [29] compared with normal controls. DTI changes in GM appear to be more sensitive than DTI changes in non-lesional WM when monitoring patients with clinically progressive disease [30].

Hypointensity of cortical and subcortical GM structures on T2-weighted images (T2-hypointesity) is another MRI manifestation of GM disease in patients with MS [31] (Fig. 3). T2-hypointensity most probably represents pathological iron deposition, as has been confirmed in studies of other neurodegenerative diseases [32]. It is not clear as yet whether this putative iron deposition is purely an epiphenomenon or whether it contributes directly to tissue injury through the generation of free radicals and lipid peroxidation [32].

Gray Matter Involvement: Clinical Relevance

Conventional MRI measures of WM damage have shown only weak to moderate correlations with physical disability, cognitive impairment, and fatigue in patients with MS [2, 9]. Initial studies indicate the clinical relevance of GM damage in helping to overcome this clinical-MRI paradox. Global GM atrophy shows stronger correlations with physical disability measures [8] (Table 1) and certain aspects of neuropsychological impairment than do conventional T1 hypointense and T2 hyperintense WM lesion and global WM atrophy measures [33]. Examination of regional GM involvement may help to uncover the structural or functional substrate underlying particular clinical manifestations of MS that have eluded conventional MRI-based WM correlation studies. For example, the involvement of specific areas of the cerebral cortex [34] and deep gray nuclei [35-38] has been linked to clinical features of MS.

Summary and Future Directions

Views concerning the involvement of GM in MS have undergone considerable evolution during the past decade. It is now clear that GM is involved to a considerable extent in MS. Such involvement begins early in the disease course and may occur, in part, independently of WM damage. GM changes assessed by MRI hold promise as new biomarkers in uncovering a component of the disease uniquely contributing to neurologic impairment and disease progression. Further studies are warranted in order to identify the mechanisms leading to GM pathology, to discover whether such mechanisms provide new therapeutic targets, to examine the relationship between GM and WM pathology, and to reveal how GM imaging adds to the search for the most sensitive and clinically relevant longitudinal biomarkers of the underlying disease process.

Acknowledgements. This work was supported by research grants to R. Bakshi from the National Institutes of Health (NIH-NINDS 1 K23 NS42379-01) and National Multiple Sclerosis Society (RG 3574A1 and RG 3705A1). We thank Ms. Sophie Tamm for her assistance with manuscript preparation.

References

1. Bagnato F, Jeffries N, Richert ND et al (2003) Evolution of T1 black holes in patients with multiple sclerosis imaged monthly for 4 years. Brain 126:1782-1789
2. Bakshi R, Minagar A, Jaisani Z, Wolinsky JS (2005) Imaging of multiple sclerosis: role in neurotherapeutics. NeuroRx 2:277-303
3. van Walderveen MA, Kamphorst W, Scheltens P et al (1998) Histopathologic correlate of hypointense lesions on T1-weighted spin-echo MRI in multiple sclerosis. Neurology 50:1282-1288
4. Bitsch A, Kuhlmann T, Stadelmann C et al (2001) A longitudinal MRI study of histopathologically defined hypointense multiple sclerosis lesions. Ann Neurol 49:793-796
5. Enzinger C, Ropele S, Smith S et al (2004) Accelerated evolution of brain atrophy and "black holes" in MS patients with APOE-epsilon 4. Ann Neurol 55:563-569
6. van Waesberghe JH, van Walderveen MA, Castelijns JA et al (1998) Patterns of lesion development in multiple sclerosis: longitudinal observations with T1-weighted spin-echo and magnetization transfer MR. Am J Neuroradiol 19:675-683
7. Truyen L, van Waesberghe JH, van Walderveen MA et al (1996) Accumulation of hypointense lesions ("black holes") on T1 spin-echo MRI correlates with disease progression in multiple sclerosis. Neurology 47:1469-1476
8. Sanfilipo MP, Benedict RH, Sharma J et al (2005) The relationship between whole brain volume and disability in multiple sclerosis: a comparison of normalized gray vs. white matter with misclassification correction. Neuroimage 26:1068-1077
9. Zivadinov R, Leist TP (2005) Clinical-magnetic resonance imaging correlations in multiple sclerosis. J Neuroimaging 15:10S-21S
10. Filippi M, Rovaris M, Rocca MA et al (2001) Glatiramer acetate reduces the proportion of new MS lesions evolving into "black holes". Neurology 57:731-733
11. Dalton CM, Miszkiel KA, Barker GJ et al (2004) Effect of natalizumab on conversion of gadolinium enhancing lesions to T1 hypointense lesions in relapsing multiple sclerosis. J Neurol 251:407-413
12. Simon JH, Lull J, Jacobs LD et al (2000) A longitudinal study of T1 hypointense lesions in relapsing MS: MSCRG trial of interferon beta-1a. Multiple Sclerosis Collaborative Research Group. Neurology 55:185-192
13. Barkhof F, van Waesberghe JH, Filippi M et al (2001) T(1) hypointense lesions in secondary progressive multiple sclerosis: effect of interferon beta-1b treatment. Brain 124:1396-1402
14. Comi G, Filippi M, Wolinsky JS (2001) European/Canadian multicenter, double-blind, randomized, placebo-controlled study of the effects of glatiramer acetate on magnetic resonance imaging-measured disease activity and burden in patients with relapsing multiple sclerosis. European/Canadian Glatiramer Acetate Study Group. Ann Neurol 49:290-297
15. Kidd D, Barkhof F, McConnell R et al (1999) Cortical lesions in multiple sclerosis. Brain 122:17-26
16. Peterson JW, Bo L, Mork S et al (2001) Transected neurites, apoptotic neurons, and reduced inflammation in cortical multiple sclerosis lesions. Ann Neurol 50:389-400
17. Roemer S, Stadelmann C, Bruck W et al (2006) Cortical demyelination is present in early multiple sclerosis. Neurology 66(Suppl 2):A93-A94
18. Bo L, Vedeler CA, Nyland HI et al (2003) Subpial demyelination in the cerebral cortex of multiple sclerosis patients. J Neuropathol Exp Neurol 62:723-732
19. Kutzelnigg A, Lucchinetti CF, Stadelmann C et al (2005) Cortical demyelination and diffuse white matter injury in multiple sclerosis. Brain 128:2705-2712
20. Bakshi R, Ariyaratana S, Benedict RH, Jacobs L (2001) Fluid-attenuated inversion recovery magnetic resonance imaging detects cortical and juxtacortical multiple sclerosis lesions. Arch Neurol 58:742-748

21. Geurts JJ, Pouwels PJ, Uitdehaag BM et al (2005) Intracortical lesions in multiple sclerosis: improved detection with 3D double inversion-recovery MR imaging. Radiology 236:254-260
22. Bermel RA, Bakshi R (2006) The measurement and clinical relevance of brain atrophy in multiple sclerosis. Lancet Neurol 5:158-170
23. Dalton CM, Chard DT, Davies GR et al (2004) Early development of multiple sclerosis is associated with progressive grey matter atrophy in patients presenting with clinically isolated syndromes. Brain 127:1101-1107
24. Valsasina P, Benedetti B, Rovaris M et al (2005) Evidence for progressive gray matter loss in patients with relapsing-remitting MS. Neurology 65:1126-1128
25. Bermel RA, Innus MD, Tjoa CW, Bakshi R (2003) Selective caudate atrophy in multiple sclerosis: a 3D MRI parcellation study. Neuroreport 14:335-339
26. Wylezinska M, Cifelli A, Jezzard P et al (2003) Thalamic neurodegeneration in relapsing-remitting multiple sclerosis. Neurology 60:1949-1954
27. Prinster A, Quarantelli M, Orefice G et al (2006) Grey matter loss in relapsing-remitting multiple sclerosis: a voxel-based morphometry study. Neuroimage 29:859-867
28. Inglese M, Liu S, Babb JS et al (2004) Three-dimensional proton spectroscopy of deep gray matter nuclei in relapsing-remitting MS. Neurology 63:170-172
29. Audoin B, Ranjeva JP, Au Duong MV et al (2004) Voxel-based analysis of MTR images: a method to locate gray matter abnormalities in patients at the earliest stage of multiple sclerosis. J Magn Reson Imaging 20:765-771
30. Rovaris M, Gallo A, Valsasina P et al (2005) Short-term accrual of gray matter pathology in patients with progressive multiple sclerosis: an in vivo study using diffusion tensor MRI. Neuroimage 24:1139-1146
31. Bakshi R, Benedict RH, Bermel RA et al (2002) T2 hypointensity in the deep gray matter of patients with multiple sclerosis: a quantitative magnetic resonance imaging study. Arch Neurol 59:62-68
32. Brass SD, Chen N, Mulkern R, Bakshi R (2006) Magnetic resonance imaging of iron deposition in neurologic disorders. Top Magn Reson Imaging 17:31-40
33. Sanfilipo MP, Benedict RH, Weinstock-Guttman B, Bakshi R (2006) Gray and white matter brain atrophy and neuropsychological impairment in multiple sclerosis. Neurology 66:685-692
34. Morgen K, Sammer G, Courtney SM et al (2006) Evidence for a direct association between cortical atrophy and cognitive impairment in relapsing-remitting MS. Neuroimage 30:891-898
35. Tjoa CW, Benedict RH, Weinstock-Guttman B et al (2005) MRI T2 hypointensity of the dentate nucleus is related to ambulatory impairment in multiple sclerosis. J Neurol Sci 234:17-24
36. Niepel G, Tench CR, Morgan PS et al (2006) Deep gray matter and fatigue in MS A T1 relaxation time study. J Neurol 253:896-902
37. Brass SD, Benedict R, Weinstock-Guttman B et al (2006) Cognitive impairment is associated with subcortical MRI gray matter T2 hypointensity in multiple sclerosis. Mult Scler 12:437-444
38. Houtchens MK, Benedict R, Killiany R et al (2005) Thalamic atrophy in multiple sclerosis: Clinical-MRI correlations. Neurology 64(Suppl 1):A260

Chapter 5

Magnetization Transfer Imaging

M. Inglese, Y. Ge, R.I. Grossman

Introduction

Although conventional MRI cannot establish the mechanisms of neurodegeneration, increasingly sophisticated imaging techniques are making it possible to study these processes in vivo. Among the quantitative MRI techniques, magnetization transfer imaging (MTI) has been the most extensively applied to the assessment of multiple sclerosis (MS), due to its sensitivity to the most destructive pathological substrates of the disease. This chapter outlines the major contributions of MTI in detecting and monitoring neurodegeneration in MS and other CNS diseases.

Basic Principles of MT-MRI

MTI refers to an application of MRI designed to explore the characteristics of non-water components in tissue [1]. Protons contained in macromolecules are relatively immobile and hence are not accessible using conventional MR sequences. However, the addition of a radiofrequency pulse that selectively saturates macromolecular protons allows an indirect assessment of this proton pool, since its normal exchange of magnetization with water-mobile protons is modified. Calculation of the effect of the MT saturation, known as magnetization transfer ratio (MTR), represents the fractional signal loss due to the complete or partial saturation of the bound proton pool, and ranges from near zero in the cerebrospinal fluid (CSF) to 50% or more in tissue with a high proportion of poorly mobile macromolecules [2].

Histopathologic Correlate of MT-MRI Changes

There are several lines of evidence suggesting that a marked reduction in MTR values in MS lesions indicates severe tissue damage [3-5]. Axonal loss is likely to be an important contributor to MTR decreases in MS for several reasons. First, in a post-mortem study, MTR reduction in both lesions and normal-appearing white matter (NAWM) was correlated with the percentage of residual axons and the degree of demyelination [5]. Second, MTR reduction has been found to correlate well with the ratio of N-acetylaspartate to creatine (an accepted marker of neuro-axonal integrity and viability) measured in MS lesions [6]. Third, low

MTR values have been found in animal models of Wallerian degeneration and diffuse axonal injury [4].

More recently, in order to determine whether demyelination and axonal loss have differential effects on MTR, Schmierer et al. [7] regressed both measures on MTR of co-registered post-mortem tissue, and found that there was a robust correlation of MTR with myelin content alone ($r = 0.74$). MTR was significantly higher in remyelinated than demyelinated lesions in two recent studies with post-mortem correlation [7]. This suggests that MTR could also be used to monitor remyelination in selected white-matter lesions during disease progression and treatment.

MT-MRI in Multiple Sclerosis

Lesions

Established MS lesions have a wide range of MTR values, with lower MTR in lesions that are hypointense on T1-weighted (T1-W) images compared with those that are isointense to NAWM [8, 9]. This suggests that low MTR is a marker of more severe, and possibly permanent, tissue damage. In a longitudinal study with monthly MT-MRI and T1-W scans, van Waesberghe et al. [9] found that MS lesions which evolved from T1-hypointense to T1-isointese when gadolinium enhancement ceased, showed a significant MTR increase, whereas a markedly decreased MTR at the time of initial enhancement was predictive of a persistent T1 hypointensity and lower MTR after 6-month follow-up. This suggests a highly variable balance between damaging and reparative mechanisms with different lesion subsets, and with different degrees of structural damage contributing to the evolution of the disease.

Normal-Appearing White Matter (NAWM)

Decreased MTR has also been found in the NAWM of MS patients, and although these changes are more pronounced in NAWM areas adjacent to focal T2-weighted (T2-W) lesions they can also be found in the absence of lesions [10]. The abnormalities seen on MTR are quantitatively small, and unlike the large changes seen in lesions, are likely to be less specific for variations in myelin content per se. There is experimental evidence that inflammation alone slightly reduces MTR [3], and axonal degeneration will also contribute to MTR decrease. MTR is perhaps the most robust quantitative measure of NAWM abnormality. Reduced MTR is already present in patients with clinically isolated syndromes (CIS) [11], increases progressively in individuals with early relapsing/remitting (RR) MS [12], and is most abnormal in patients with secondary progressive (SP) disease [10]. MT-MRI can also be used to assess global MS tissue burden by means of MTR histogram analysis [13]. This is a highly automated technique capable of providing several metrics reflecting both macroscopic and microscopic MS pathology. MS patients typically have lower average MTR, and lower histogram peak height and position, than normal subjects. MTR histogram parameters can be different in the various clinical forms of MS [14]. Primary progressive MS (PPMS) patients have significantly lower histogram peak height with normal peak position and only slightly reduced average MTR, suggesting the presence of a subtle but widespread damage of the NAWM. RRMS patients have lower average MTR and peak height than benign MS patients, whose histograms are similar to those of healthy individuals. Patients with SPMS had the lowest MTR histogram metrics. In addition, in a cohort of 73 MS patients with various clinical subtypes, whole brain MTR histograms were related to development of disability over the next 4.5 years [15].

Normal-Appearing Gray Matter (NAGM)

It is noteworthy that, although axonal damage has been reported in classical neuropathology for over a century, only recently have post-mortem studies shown that neurons and axons are also targets of the MS process [16] and that their dysfunction may contribute to clinical disability [17]. Indeed, Peterson at al. [18] reported that cerebral cortex MS lesions are characterized by demyelination, axonal and dendritic transaction, and apoptotic loss of neurons. In addition, subcortical lesions which result in axonal transection might also be accompanied by retrograde neuronal degeneration. Unfortunately, such lesions are frequently missed by conventional T2-W images, due to their small size, poor contrast with surrounding gray matter (GM), and partial volume effects with adjacent WM. However, recent studies have shown reduced MTR values in the GM from patients with MS, using region of interest (ROI) [19] or histogram [19, 20] analysis (see Fig. 1). Interestingly, in

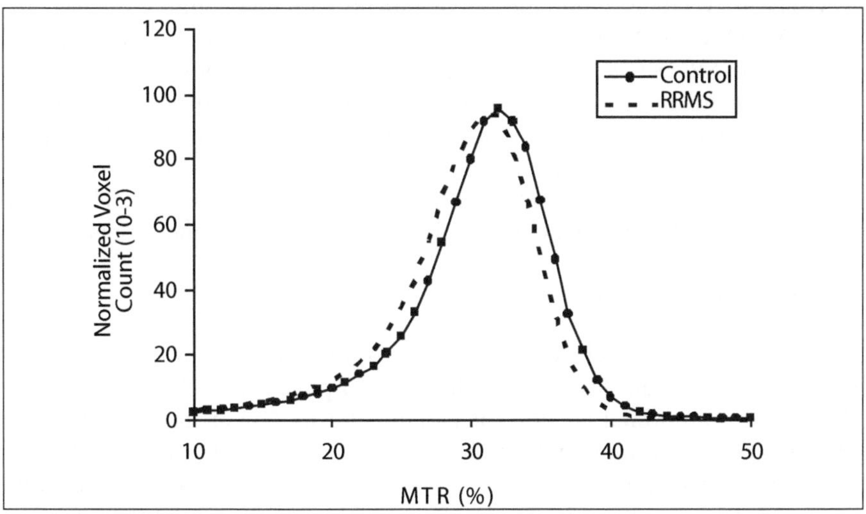

Fig. 1. Average gray matter MTR histograms from relapsing/remitting MS patients (dotted line) and age-matched healthy controls (solid line). Note that, although the peak height is similar in the two groups of subjects, the whole body of the histogram is shifted to the left in MS patients

one study, the average percentage reduction of the peak height of the GM-MTR histogram from MS patients was of the same magnitude (about 20%) as the average percentage reduction of the peak height of the cortical GM-MTR histograms from patients with Alzheimer's disease (AD) [21]; and in another [20], the peak height of the GM-MTR histogram was inversely correlated with the severity of clinical disability ($r = -0.65$). Significant correlations have also been reported between GM-MTR histogram-derived changes and T2 lesion volume [19, 20]. This fits with the notion that at least part of GM pathology in MS is secondary to retrograde degeneration of fibers traversing white-matter lesions. Very importantly, reduced GM MTR has also been seen in CIS patients [11], suggesting that early GM abnormalities are more common than would be expected from neuropathological studies, which are more likely to sample patients with late progressive disease [22]. In addition, in a cohort of 73 MS patients with various clinical subtypes, GM MTR histograms were related to development of disability over the next 8 years [23].

MT-MRI of Spinal Cord

Much of the locomotor disability in MS comes from spinal cord pathology but, as in the brain, there is a poor correlation between the number of spinal cord T2 lesions and disability [24]. On the other hand, average cervical cord MTR is lower in MS patients with locomotor disability than in those without [25]. Postmortem studies have shown both focal and diffuse regions of demyelination and axonal loss in MS spinal cord, with involvement of both GM [26, 27] and WM. Diffuse hyperintensity may be seen on proton density-weighted images of the cord, especially in primary progressive (PP) MS. A recent study has compared cervical cord MTR histogram metrics of patients with PPMS and secondary progressive (SP) MS and found no significant difference between these two groups, even though SPMS patients had higher MRI-visible lesion burdens [28]. These data indicate the presence of diffuse tissue damage undetectable by conventional MRI in PPMS patients, the extent of which seems to match that of SPMS patients with similar levels of disability. While routine T2-weighted images detect foci of demyelination, they are insensitive to the extent of axonal loss [29]. The global measures of MTR may more closely reflect diffuse pathology and axonal loss.

MT-MRI of the Optic Nerve

Optic nerve imaging is challenging because of the small size of the nerve, confounding signals from the surrounding cerebrospinal fluid (CSF) in the nerve sheath and from orbital fat, and the potential for motion artifacts. Optic nerve MTR has been measured on 2D and 3D sequences [30], and a correlation made with visual evoked potential latency [31] and with the degree of visual function recovery after an acute episode of optic neuritis [30]. In a recent study, optic nerve (ON) volume and MTR from MS patients who had incomplete or no visual recovery after optic neuritis were compared with those from MS patients who showed a marked clinical recovery, and with those from patients with Leber hereditary optic neuropathy (LHON), whose major histopathological hallmark of impaired vision is neuronal and axonal loss [30]. Interestingly, volumes and MTR values of the ONs from patients with MS and incomplete or no recovery were lower than those of the ONs from patients with MS and recovery, but no different from those of patients with LHON, thus suggesting that markedly low MTR may be a marker of permanent tissue damage.

MT-MRI of Other Neurological Diseases

Although MT-MRI has been mainly applied to the study of MS and other white-matter diseases, brain pathology of patients with neurodegenerative diseases has also been quantified using this technique. Patients with Alzheimer's disease (AD) have a markedly reduced MTR histogram peak height of the cortical GM and temporal lobe GM [21]. In these patients, a composite MR score based on brain volume and cortical GM-MTR histogram peak height was correlated with patients' cognitive impairment ($r = 0.65$). These results have been confirmed by other two studies [32, 33], which also showed decreased MTR values of the cortical GM and temporal lobe GM from patients with mild cognitive impairment (MCI), in the absence of significant volumetric changes. This suggests that MTR has the potential to detect the progression of MCI to AD before CNS atrophy is seen on conventional MRI scans. Patients with AD have also been found to have reduced MTR values of the hippocampus [34]. Interestingly, hippocampus MTR yielded a relatively high discrimination rate between the very mild AD group and the control group, with a sensitivity of 75% and a specificity of 90%, resulting in an overall discrimination rate of 85%. This suggests that MTR values may be useful for the early detection of AD. Reduced MTR values have also been found in the cortical spinal tracts of patients with amyotrophic lateral sclerosis [35]. This has led to the speculation that MTR decrease reflects pyramidal tract degeneration and that MTR may provide objective and quantitative information on the severity of neurodegeneration.

Conclusions

The evidence of a relevant neurodegenerative component in MS pathology has prompted the need for neuroprotective treatments and imaging markers able to quantify neurodegeneration and its response to current and experimental treatments. MTI can provide relevant information on the structural changes occurring within and outside macroscopic lesions. The application of this technique in patients with MS and other neurodegenerative diseases should improve the understanding of the mechanisms leading to irreversible disability and improve our ability to monitor the effect of new treatments in clinical neurology.

Acknowledgements. Our work is supported by the National Institute of Health (NIH) grants R37 NS 29029-11 and RO1 NS051623-01.

References

1. Balaban RS, Ceckler TL (1992) Magnetization transfer contrast in magnetic resonance imaging. Magn Reson Q 8:116-137
2. Henkelman RM, Huang X, Xiang QS et al (1993) Quantitative interpretation of magnetization transfer. Magn Res Med 29:759-766
3. Dousset V, Grossman RI, Ramer KN et al (1992) Experimental allergic encephalomyelitis and multiple sclerosis: lesion characterization with magnetization transfer imaging [published erratum appears in Radiology 1992 Jun;183(3):878]. Radiology 182:483-491
4. Lexa FJ, Grossman RI, Rosenquist AC (1994) Dyke Award paper. MR of Wallerian degeneration in the feline visual system: characterization by magnetization transfer rate with histopathologic correlation. Am J Neuroradiol 15:201-212
5. van Waesberghe JH, Kamphorst W, De Groot CJ et al (1999) Axonal loss in multiple sclerosis lesions: magnetic resonance imaging insights into substrates of disability. Ann Neurol 46:747-754
6. Hiehle JF, Jr, Lenkinski RE, Grossman RI et al (1994) Correlation of spectroscopy and magnetization transfer imaging in the evaluation of demyelinating lesions and normal appearing white matter in multiple sclerosis. Magn Res Med 32:285-293
7. Schmierer K, Scaravilli F, Altmann DR et al (2004) Magnetization transfer ratio and myelin in postmortem multiple sclerosis brain. Ann Neurol 56:407-415
8. Hiehle JF, Jr, Grossman RI, Ramer KN et al (1995) Magnetization transfer effects in MR-detected multiple sclerosis lesions: comparison with gadolinium-enhanced spin-echo images and nonenhanced T1-weighted images. Am J Neuroradiol 16:69-77
9. van Waesberghe JH, van Walderveen MA, Castelijns JA et al (1998) Patterns of lesion development in multiple sclerosis: longitudinal observations with T1-weighted spin-echo and magnetization transfer MR. Am J Neuroradiol 19:675-683
10. Filippi M, Campi A, Dousset V et al (1995) A magnetization transfer imaging study of normal-appearing white matter in multiple sclerosis. Neurology 45:478-482
11. Fernando KT, Tozer DJ, Miszkiel KA et al (2005) Magnetization transfer histograms in clinically isolated syndromes suggestive of multiple sclerosis. Brain 128:2911-2925
12. Davies GR, Altmann DR, Hadjiprocopis A et al (2005 Increasing normal-appearing grey and white matter magnetisation transfer ratio abnormality in early relapsing-remitting multiple sclerosis. J Neurol 252:1037-1044
13. van Buchem MA, McGowan JC, Kolson DL et al (1996) Quantitative volumetric magnetization transfer analysis in multiple sclerosis: estimation of macroscopic and microscopic disease burden. Magn Res Med 36:632-636
14. Filippi M, Iannucci G, Tortorella C et al (1999) Comparison of MS clinical phenotypes using conventional and magnetization transfer MRI. Neurology 52:588-594
15. Rovaris M, Agosta F, Sormani MP et al (2003) Conventional and magnetization transfer MRI predictors of clinical multiple sclerosis evolution: a medium-term follow-up study. Brain 126:2323-2332
16. Trapp BD, Peterson J, Ransohoff RM et al (1998) Axonal transection in the lesions of multiple sclerosis. N Eng J Med 338:278-285
17. Bjartmar C, Trapp BD (2003) Axonal degeneration and progressive neurologic disability in multiple sclerosis. Neurotox Res 5:157-164
18. Peterson JW, Bo L, Mork S et al (2001) Transected neurites, apoptotic neurons, and reduced inflammation in cortical multiple sclerosis lesions. Ann Neurol 50:389-400
19. Cercignani M, Bozzali M, Iannucci G et al (2001) Magnetisation transfer ratio and mean diffusivity of normal appearing white and grey matter from patients with multiple sclerosis. J Neurol Neurosurg Psychiat 70:311-317

20. Ge Y, Grossman RI, Udupa JK et al (2001) Magnetization transfer ratio histogram analysis of gray matter in relapsing-remitting multiple sclerosis. Am J Neuroradiol 22:470-475
21. Bozzali M, Franceschi M, Falini A et al (2001) Quantification of tissue damage in AD using diffusion tensor and magnetization transfer MRI. Neurology 57:1135-1137
22. Kutzelnigg A, Lucchinetti CF, Stadelmann C et al (2005) Cortical demyelination and diffuse white matter injury in multiple sclerosis. Brain 128:2705-2712
23. Agosta F, Rovaris M, Pagani E et al (2006) Magnetization transfer MRI metrics predict the accumulation of disability 8 years later in patients with multiple sclerosis. Brain 129:2620-2627
24. Kidd D, Thorpe JW, Thompson AJ et al (1993) Spinal cord MRI using multi-array coils and fast spin echo. II. Findings in multiple sclerosis. Neurology 43:2632-2637
25. Filippi M, Bozzali M, Horsfield MA et al (2000) A conventional and magnetization transfer MRI study of the cervical cord in patients with MS. Neurology 54:207-213
26. Mottershead JP, Schmierer K, Clemence M et al (2003) High field MRI correlates of myelin content and axonal density in multiple sclerosis: a post-mortem study of the spinal cord. J Neurol 250:1293-1301
27. Gilmore CP, Bo L, Owens T et al (2006) Spinal cord gray matter demyelination in multiple sclerosis: a novel pattern of residual plaque morphology. Brain Pathol 16:202-208
28. Rovaris M, Bozzali M, Santuccio G et al (2001) In vivo assessment of the brain and cervical cord pathology of patients with primary progressive multiple sclerosis. Brain 124:2540-2549
29. Bot JC, Blezer EL, Kamphorst W et al (2004) The spinal cord in multiple sclerosis: relationship of high-spatial-resolution quantitative MR imaging findings to histopathologic results. Radiology 233:531-540
30. Inglese M, Ghezzi A, Bianchi S et al (2002) Irreversible disability and tissue loss in multiple sclerosis: a conventional and magnetization transfer magnetic resonance imaging study of the optic nerves. Arch Neurol 59:250-255
31. Hickman SJ, Toosy AT, Jones SJ et al (2004) Serial magnetization transfer imaging in acute optic neuritis. Brain 127:692-700
32. Kabani NJ, Sled JG, Chertkow H (2002) Magnetization transfer ratio in mild cognitive impairment and dementia of Alzheimer's type. Neuroimage 15:604-610
33. Kabani NJ, Sled JG, Shuper A, Chertkow H (2002) Regional magnetization transfer ratio changes in mild cognitive impairment. Magn Res Med 47:143-148
34. Hanyu H, Asano T, Sakurai H et al (2001) Magnetization transfer measurements of the hippocampus in the early diagnosis of Alzheimer's disease. J Neurol Sci 188:79-84
35. Kato Y, Matsumura K, Kinosada Y et al (1997) Detection of pyramidal tract lesions in amyotrophic lateral sclerosis with magnetization-transfer measurements. Am J Neuroradiol 18:1541-1547

Chapter 6

Perfusion MRI

Y. Ge, M. Law, M. Inglese, R.I. Grossman

Introduction

There has been an increasing interest in studying microvascular and brain perfusion abnormalities in multiple sclerosis (MS) due to the accumulating evidence regarding the primary vascular pathogenesis in MS [1-4]. Early microscopic investigations demonstrated perivascular inflammatory infiltrates, including macrophages, often found in close contact with the disintegrating myelin sheath [5], which were later described as a T-cell-mediated immune reaction causing the activation of macrophages that contained intracytoplasmic, myelin-reactive degradation products [6]. The intravascular fibrin deposition [7, 8] and venous occlusive changes with hemodynamic impairment have also been demonstrated in active MS lesions [9]. Therefore, not surprisingly, these essential vascular pathologies have effects on the cerebral blood perfusion, which may cause mitochondrial dysfunction and axonal degeneration. However, only recently, due to the major achievements of in vivo perfusion imaging, have we started to understand the mechanisms of hemodynamic changes and neurodegeneration occurring in MS. This chapter focuses on some of the recent developments in the field of vascular pathology and on some ongoing research using perfusion MRI, particularly dynamic susceptibility contrast-enhanced MRI (DSC-MRI), to investigate vascular neuropathology in MS.

Vascular Pathology and Ischemic Hypothesis

Lesion Heterogeneity

The perivascular (mostly perivenular) inflammation, in which the cell infiltration from blood to parenchyma is mainly composed of lymphocytes and macrophages, has long been described as a critical event in tissue injury in MS [2]. The T-lymphocyte-mediated inflammation can be associated with a zone of demyelination which surrounds the venules in the early stages of the disease. However, the inflammatory basis of demyelination during lesion formation is not constant in all acute MS lesions; recent histopathologic studies have suggested that MS is best characterized as a heterogeneous disease. By studying over 200 MS cases with biopsy and autopsy examination, Lucchinetti et al. [10, 11] described four patterns of MS lesion pathology, which include T-cell/macrophage toxins (Pattern I), antibody/complement associated (Pattern II), distal oligodendrogliopathy (Pattern III), and oligodendrocyte degeneration in the periplaque white matter (Pattern IV). This is consistent with the heterogeneity, a striking feature in MS, with respect to the clinical course, neuroradiological appearance, and response to therapy. The different patterns suggest different pathways to myelin injury and the underlying pathology, which in turn can be either due to direct inflammation cell-mediated mechanisms accompanied by inflammatory activators such as macrophages or antibodies (Patterns I and II) or as a result of primary or secondary dysfunction of oligodendrocytes (Patterns III and IV). This will, at least, lead to two histopathologic scenarios: one in which the remyelinated shadow plaques are common, and another in which the remyelination is absent [12].

Pattern-III demyelination is defined as distal "dying-back" oligodendrogliopathy [11], in which the typical findings are preferential loss of myelin-associated glycoprotein, apoptotic oligodendrocytes resulting in limited remyelination, and ill-defined inflammatory lesion borders. Pattern III has attracted particular interest in the search for a better understanding of the pathogenetic heterogeneity of MS, because this characteristic disturbance of oligodendrocytes closely mimics the tissue pathology in the early stages of white-matter ischemia, suggesting a hypoxic white-matter injury as a pathogenetic component in a subset of MS patients [4]. Barnett and Prineas [13] have confirmed this pattern-III tissue injury in a group of patients who died shortly after the onset of a relapse, by showing an extensive oligodendrocyte apoptosis in the absence of inflammatory cells, indicating an abnormality in the local environment, to which oligodendrocytes are susceptible, initiating apoptosis. This has not been seen in any experimental allergic encephalomyelitis model but is reported in hypoxic stress secondary to ischemia or an immune-mediated vasculitis [2]. The authors conclude that this probably represents a very early stage in most, if not all, MS lesions. Although it is well recognized that there is vascular injury in MS, it has attracted particular attention recently by showing hypoxia-like tissue injury in histopathologic studies of MS [4].

Inflammation and Thrombosis

There are several mechanisms by which a primary vascular pathology could lead to disease progression in MS. Given the close histopathologic relationship between MS lesions and cerebral veins [14] and the fact that central veins are almost invariably found in MS lesions on MR venography [15], it has become increasingly clear that vascular inflammatory pathology could represent an early phase in the formation of acute enhancing lesions. The luminal endothelial surface becomes actively prothrombotic under the influence of inflammatory mediators, which not only induces microvascular procoagulant activity but also expresses platelet-activating factor and inhibits fibrinolysis [16], leading to microvascular fibrin deposition and thrombosis. The histopathologic evidence of vascular occlusion was described in early histopathologic studies [17] showing vascular inflammation contributing to vascular thrombosis. Using immunohistochemical techniques, Wakefield et al. [3] later confirmed these findings in three acute cases of MS, demonstrating fibrin deposition and thrombosis of vessels associated with endothelial cell activation, suggesting that thrombosis of small veins and capillaries could represent an ischemic basis for disease. These vascular changes were seen prior to cerebral parenchymal reaction and demyelination, and were not seen in control cerebral tissues. Adams [18] reviewed tissue from 70 cases of MS and found evidence of vascular (venous) damage in 41%, including fibrinoid deposition, hemosiderin, thrombosis, and thickened veins. Occlusive changes have also been found clinically in the retinal venules of patients who later developed MS [9].

Despite the role of inflammation in vascular occlusive changes, it is still not known whether inflammation or thrombosis is the primary insult in MS. Researchers in stroke studies have demonstrated that experimental ischemia can induce a massive inflammatory response through upregulating proinflammatory genes and activating transcription factors [19, 20]. These factors promote expression of adhesion molecules and allow infiltration of inflammatory cells including macrophages and T-cells into the brain, thereby promoting further injury including neurodegeneration [21]. Thus, taken together with the hypoxia-like tissue injury found recently, it seems possible that a primary ischemic pathology could be an initial event in MS, which leads to a subsequent inflammatory reaction in MS.

Mitochondrial Dysfunction and Energy Failure

Neurons are susceptible to ischemic injury because of their low tolerance to local oxygen deprivation. In the deep watershed white-matter region, the relatively lesser and slower blood flow often leads to ischemic lesions [22]. Cerebral perfusion decrease is not only found in stroke, but also in many neurodegenerative diseases. In ischemic injury, failure of the Na-K-ATPase pump, due to inadequate delivery of oxygen and glucose, leads to Ca2+ overload, which can cause a fatal cascade of Ca2+-mediated transient or permanent injury of neurons. In MS, although endothelial damage and microvascular thrombosis are common, there is no proven significant association between vascular pathology or occlusive changes and hypoxia-like tissue injury [23]. However, mitochondria are the most vulnerable cellular organelles and are highly susceptible to hypoxia/ischemia. Oxidative damage of mitochondria DNA can be seen in both active and chronic MS lesions, with a subsequent reduction in the activity of mitochondrial enzymes [24]. In addition, many inflammatory mediators and products, such as reactive oxygen and nitric oxide intermediates, can provoke mitochondrial dysfunction by causing ATP depletion, leading to axonal injury even in the absence of demyelination [25]. Current evidence indicates that mitochondrial impairment is a universal contributor to neurodegeneration. What is fascinating is the similarity between MS lesions and acute white-matter stroke, hypothesizing a hypoxia-like injury in MS, as pointed out by Lassmann [4]. Although the ischemic hypothesis may still be best described as a working model in MS, assessing changes in cerebral perfusion due to vascular inflammation and hypoxia is intriguing. However, such microcirculation abnormalities were difficult to detect and appreciate until the application of perfusion imaging technology.

Dynamic Susceptibility Contrast-Enhanced Perfusion MRI

Dynamic susceptibility contrast-enhanced MRI (DSC-MRI) is a clinically proven perfusion method and has been validated and utilized in routine clinical practice in many institutions. DSC-MRI is based on the acquisition of a series of images during the transit of a bolus of gadolinium-based contrast material through the brain vasculature. Cerebral perfusion can be visualized by the dynamic fast imaging (typically echo planar imaging) with a temporal resolution of 1 or 2 seconds. DSC-MRI provides comprehensive neurovascular information by producing parametric perfusion maps including cerebral blood flow (CBF), cerebral blood volume (CBV), and mean transit time (MTT). Perfusion assessment adds novel pathophysiologically relevant information in a variety of situations, such as acute cerebral ischemia, tumors, and chronic vascular disease. Arterial spin labeling (ASL), another perfusion MRI technique, which does not use extrinsic contrast agents, applies a spatially selective inversion pulse to invert the spins that flow into tissue, causing a change within the tissue which is proportional to perfusion. Compared to ASL, DSC-MRI has much higher spatial resolution and can measure CBF, CBV, and MTT simultaneously, while ASL can only quantify CBF. In addition, DSC-MRI allows a larger coverage of the brain in one acquisition and uses a much shorter scanning time. All these features ensure that DSC-MRI is gaining more favor in clinical practice for measuring blood perfusion without the radiation effects that are found when using positron-emission tomography or single photon emission computed tomography (SPECT). Since patients with MS are typically undergoing gadolinium contrast injection, the addition of the DSC-MRI adds essentially no risk or expense beyond that of the routine MR examination.

The quantification of perfusion using DSC-MRI is based on intravascular indicator dilution theory [26], which is applied straightforwardly to show that the area under the concentration-time curve is proportional to CBV. Since, however, the proportionality constant is not known, this value is expressed relative to a tissue standard, typically the contralateral white matter. The measurement is then known as relative CBV (rCBV). The drawback is that this relative measurement is not appropriate in diseases that affect the brain globally, including the contralateral side. However, this approach assumes that the contrast agent remains intravascular and makes no estimate of vascular permeability, a parameter that may allow characterization of blood-brain barrier (BBB) compromise in lesions. In MS, this is generally not the case, since BBB can be disrupted in some acute lesions and the contralateral normal-appearing white matter (NAWM) is actually abnormal. By measuring the concentration of contrast in the blood (intravascular space) and in the lesion (extravascular space), it is possible to measure vascular permeability and make corrections for blood perfusion measurements [27]. The algorithm does not assume an intact BBB and avoids the overestimation of CBV that occurs with an approach based on intravascular indicator dilution theory when the BBB is actually damaged. A gamma variate function fit can be applied to partially correct for these effects. With further modification of this method using arterial input function (AIF) [28], CBV, CBF, and MTT can be quantified in a more absolute and direct manner. The AIF can be found using either an automated or a manual method [28]. Although the contrast bolus is not instantaneous, an approximation to the idealized response can be found by deconvolving the measured tissue concentration with the AIF.

Insights into Hemodynamic Abnormalities in MS Using DSC-MRI

Normal-Appearing White Matter

A limited number of studies have examined the DSC-MRI of MS and little is known about the hemodynamic changes in MS [29-31]. Law et al. [32] demonstrated a significantly decreased CBF and prolonged MTT in periventricular regions of NAWM in patients with MS compared with controls. This occurred in the setting of normal CBV, suggesting that the degree of hypoperfusion was not severe enough to produce either infarction or significant ischemia. Adhya et al. [33] showed a decrease in CBF and CBV in several regions of NAWM in both primary progressive (PP) and relapsing/remitting (RR) MS patients. Compared with RRMS patients, PPMS patients showed significantly lower CBF in the periventricular NAWM and lower CBV in the periventricular and frontal NAWM [33]. Indeed, CBV can be either elevated or decreased in stroke, depending on the level of hypoperfusion and collateral or autoregulated blood flow [34, 35]. It is likely that a modest reduction in CBF would stimulate autoregulatory mechanisms such as vasodilatation and enhancement of the collateral circulation, which would tend to increase CBV and further prolong MTT [36]. Since cerebral ischemia can induce local inflammatory processes, as in other inflammatory diseases, inflamed vessels cause vasodilatation, leading to increased perfusion (CBV). It is also noted that in the periventricular NAWM, where lesions are more prone to occur, there was significantly higher CBF and CBV than intermediate and subcortical NAWM at the level of the lateral ventricle, further supporting a vascular pathogenesis in MS. Using an ASL technique, Rashid et al. [37] found increased perfusion in some NAWM regions in patients with MS. These inconsistent findings may be due to not only the different techniques applied in these studies but also the high variability of normal blood perfusion in white matter [32].

Lesions

Studies have shown that most MS lesions demonstrate hypoperfusion characterized by decreased CBF and CBV [29, 30], which is consistent with the venous occlusive changes in MS lesions observed in histopathologic studies [3, 17]. Although hypoperfusion in MS lesions has been identified, it is unknown when these perfusion deficits start to occur and how they progress. The chronic plaques lack perivenular inflammation, but often contain evidence of vascular injury such as thickening or hyalinization of vein walls [8]. MS is a white-matter disease and white-matter ischemia is exacerbated by the microvascular abnormalities and marginal blood supply compared with gray matter. The tissue injury related to ischemia, which is associated with the histopathological type of pattern-III demyelination, may occur before inflammatory events. However, lesions formed with this mechanism may not have typical enhancement such as those with an autoimmune basis in the initial developmental stage. Lucchinetti et al. [10] have indicated that the sharp border at the active lesions with acute inflammatory basis is strictly associated with the presence of ring enhancement and hypointense T2 rims, whereas these imaging features are not found in pattern-III lesions. In MS, such tissue injury with an ischemic basis can occur independently of inflammation or other parenchymal abnormalities [13].

In MS lesions, the extent of vascular injury can be assessed with gadolinium-enhanced T1-weighted imaging. The acute inflammatory lesions show enhancement due to a local disrupted BBB, which is considered as the initial event ultimately responsible for lesion formation and axonal transection. Recently, microvascular abnormalities in MS lesions have been further appreciated with the utilization of perfusion MR technology. Several perfusion studies have shown increased blood perfusion in acute MS lesions [29-31] relative to contralateral NAWM, suggesting that a local inflammation induced vasodilatation in the brain. This is because the vasculitis-mediated injury within vessel walls can stimulate perivascular cells to secrete a number of vasoactive factors, which could modulate vascular tone and CBF [38]. However, it is still unknown how this perivascular inflammation is initiated and triggered. Studies of stroke have suggested that experimental cerebral ischemia can induce local inflammatory processes [20, 39], which in turn may cause vasodilatation and increased perfusion.

Perfusion studies of both the clinical [31] and animal models of MS [40] have shown that local perfusion changes (i.e., CBV) preceded overt changes in BBB permeability, T2 signal, and diffusion imaging metrics, suggesting that perfusion changes may be a more sensitive indicator of lesion onset than conventional indices. Non-enhancing lesions often have thickened hyalined vein walls and are less likely to demonstrate perivenular inflammation; however, new inflammatory activity can be found around these chronic lesions. Interestingly, not all non-enhancing lesions have hypoperfusion; Ge et al. [30] have shown that some non-enhancing lesions also have increased perfusion, as seen in enhancing lesions, reflecting renewed inflammatory activity in these lesions, which are not detected on conventional MRI. These observations indicate that DSC-MRI has the potential to identify the new or renewed lesion inflammatory activity before BBB disruption, potentially a marker of early vascular activity.

Gray Matter

A study using DSC-MRI to study deep gray-matter nuclei, including putamen and thalamus, has shown significantly decreased perfusion in patients with RRMS compared with normal controls [41]. Using ASL, Rashid et al. [37] also demonstrated decreased perfusion in cortical gray matter, which is apparent in primary and secondary progressive MS. These observations are consistent with an early study using SPECT [42], in which a significantly reduced regional CBF was found in frontal gray matter of progressive, but not RRMS, patients. N-acetylaspartate (NAA), which is synthesized in the mitochondria, is almost exclusively found in neuronal cells and their processes. A recent report [43] has indicated a 3.6-fold disparity between NAA loss (faster) and brain parenchyma loss, indicating that neuronal/axonal loss is not the only cause of NAA reduction; mitochondrial dysfunction on a mild diffuse ischemic basis (rather than neuronal loss) may therefore play an important role in this context. On metabolic imaging with MRS, a high concentration of lactate observed in acute demyelinating lesions [44], along with decreased NAA and restricted diffusion [45], may possibly suggest an ischemic injury.

The hypoperfusion in the deep gray matter may also contribute to the increased iron deposition in these regions with iron-enriched nucleus [46]. T2 shortening, a result of nonheme iron deposition, has been reported in the basal ganglia and in the cortical regions [47] and can be associated with brain atrophy. Excessive unconjugated ferrous iron is toxic to neuronal cells, mediated through the production of hydroxide radicals [48]. The oxidative damage produced by iron has been implicated in the injury of both oligodendrocytes and neuronal cells [49]. Therefore, the study of hemodynamic impairment and abnormal iron accumulation may have clinical implications for neuroprotective therapy and improved treatment strategies.

Conclusions

Cerebral perfusion is critical in maintaining the normal brain function, and altered perfusion is associated with abnormal parenchymal metabolism and the extent of neuronal dysfunction. Perfusion data complements conventional MRI and provides comprehensive neurovascular examination. Perfusion studies indicate that vascular hemodynamic abnormality is a significant component in the pathophysiology of MS which may lead to neurodegenerative changes. Assessment of cerebral blood perfusion might not only provide a more sensitive marker of tissue ischemia, but may also bring new insights into the inflammation-mediated vascular pathology occurring in MS, which in turn may offer new targets for intervention.

Acknowledgements. Our work is supported by National Institute of Health (NIH) grants R37 NS 29029-11 and RO1 NS051623-01

References

1. Putnam TJ (1933) The pathogenesis of multiple sclerosis: a possible vascular factor. N Engl J Med 209:786-790
2. Adams CW, Abdulla YH, Torres EM, Poston RN (1987) Periventricular lesions in multiple sclerosis: their perivenous origin and relationship to granular ependymitis. Neuropathol Appl Neurobiol 13:141-152
3. Wakefield AJ, More LJ, Difford J, McLaughlin JE (1994) Immunohistochemical study of vascular injury in acute multiple sclerosis. J Clin Pathol 47:129-133
4. Lassmann H (2003) Hypoxia-like tissue injury as a component of multiple sclerosis lesions. J Neurol Sci 206:187-191
5. Charcot J-M (1877) Lectures on the diseases of the nervous system. New Sydenham Society, London
6. Traugott U, Reinherz EL, Raine CS (1983) Multiple sclerosis: distribution of T cell subsets within active chronic lesions. Science 219:308-310
7. Putnam TJ (1937) Lesions of 'encephalomyelitis' and multiple sclerosis: venous thrombosis as the primary alteration. JAMA 108:1477-1480
8. Adams CW, High OB (1982) Embolism and multiple sclerosis. Lancet 1:621
9. Graham EM, Stanford MR, Sanders MD, Kasp E, Dumonde DC (1989) A point prevalence study of 150 patients with idiopathic retinal vasculitis. 1. Diagnostic value of ophthalmological features. Br J Ophthalmol 73:714-721
10. Lucchinetti CF, Bruck W, Lassmann H (2004) Evidence for pathogenic heterogeneity in multiple sclerosis. Ann Neurol 56:308
11. Lucchinetti C, Bruck W, Parisi J et al (2000) Heterogeneity of multiple sclerosis lesions: implications for the pathogenesis of demyelination. Ann Neurol 47:707-717
12. Lassmann H, Bruck W, Lucchinetti C (2001) Heterogeneity of multiple sclerosis pathogenesis: implications for diagnosis and therapy. Trends Mol Med 7:115-121
13. Barnett MH, Prineas JW (2004) Relapsing and remitting multiple sclerosis: pathology of the newly forming lesion. Ann Neurol 55:458-468
14. Fog T (1964) On the vessel-plaque relationships in the brain in multiple sclerosis. Acta Neurol Scand 40(Suppl 10):9-15
15. Tan IL, van Schijndel RA, Pouwels PJ et al (2000) MR venography of multiple sclerosis. Am J Neuroradiol 21:1039-1042
16. Libby P, Ordovas JM, Auger KR et al (1986) Endotoxin and tumor necrosis factor induce interleukin-1 gene expression in adult human vascular endothelial cells. Am J Pathol 124:179-185
17. Putnam TJ (1935) Evidences of vascular occlusion in multiple sclerosis and encephalomyelitis. Arch Neurol Neuropsychol 32:1298-1321
18. Adams CW (1988) Perivascular iron deposition and other vascular damage in multiple sclerosis. J Neurol Neurosurg Psychiat 51:260-265

19. Wang X, Barone FC, Aiyar NV, Feuerstein GZ (1997) Interleukin-1 receptor and receptor antagonist gene expression after focal stroke in rats. Stroke 28:155-161; comment 161-162
20. del Zoppo GJ, Becker KJ, Hallenbeck JM (2001) Inflammation after stroke: is it harmful? Arch Neurol 58:669-672
21. Barone FC, Feuerstein GZ (1999) Inflammatory mediators and stroke: new opportunities for novel therapeutics. J Cereb Blood Flow Metab 19:819-834
22. Andeweg J (1996) The anatomy of collateral venous flow from the brain and its value in aetiological interpretation of intracranial pathology. Neuroradiology 38:621-628
23. Aboul-Enein F, Lassmann H (2005) Mitochondrial damage and histotoxic hypoxia: a pathway of tissue injury in inflammatory brain disease? Acta Neuropathol (Berl) 109:49-55
24. Lu F, Selak M, O'Connor J et al (2000) Oxidative damage to mitochondrial DNA and activity of mitochondrial enzymes in chronic active lesions of multiple sclerosis. J Neurol Sci 177:95-103
25. Aboul-Enein F, Weiser P, Hoftberger R et al (2006) Transient axonal injury in the absence of demyelination: a correlate of clinical disease in acute experimental autoimmune encephalomyelitis. Acta Neuropathol (Berl) 111:539-547
26. Rosen BR, Belliveau JW, Vevea JM, Brady TJ (1990) Perfusion imaging with NMR contrast agents. Magn Reson Med 14:249-265
27. Johnson G, Wetzel SG, Cha S et al (2004) Measuring blood volume and vascular transfer constant from dynamic, T*2-weighted contrast-enhanced MRI. Magn Reson Med 51:961-968
28. Carroll TJ, Rowley HA, Haughton VM (2003) Automatic calculation of the arterial input function for cerebral perfusion imaging with MR imaging. Radiology 227:593-600
29. Haselhorst R, Kappos L, Bilecen D et al (2000) Dynamic susceptibility contrast MR imaging of plaque development in multiple sclerosis: application of an extended blood-brain barrier leakage correction. J Magn Reson Imaging 11:495-505
30. Ge Y, Law M, Johnson G et al (2005) Dynamic susceptibility contrast perfusion MR imaging of multiple sclerosis lesions: characterizing hemodynamic impairment and inflammatory activity. Am J Neuroradiol 26:1539-1547
31. Wuerfel J, Bellmann-Strobl J, Brunecker P et al (2004) Changes in cerebral perfusion precede plaque formation in multiple sclerosis: a longitudinal perfusion MRI study. Brain 127:111-119
32. Law M, Saindane AM, Ge Y et al (2004) Microvascular abnormality in relapsing-remitting multiple sclerosis: perfusion MR imaging findings in normal-appearing white matter. Radiology 231:645-652
33. Adhya S, Johnson G, Herbert J et al (2006) Pattern of hemodynamic impairment in multiple sclerosis: Dynamic susceptibility contrast perfusion MR imaging at 3.0 T. Neuroimage 33:1029-1035
34. Karonen JO, Vanninen RL, Liu Y et al (1999) Combined diffusion and perfusion MRI with correlation to single-photon emission CT in acute ischemic stroke. Ischemic penumbra predicts infarct growth. Stroke 30:1583-1590
35. Hatazawa J, Shimosegawa E, Toyoshima H et al (1999) Cerebral blood volume in acute brain infarction: A combined study with dynamic susceptibility contrast MRI and 99mTc-HMPAO-SPECT. Stroke 30:800-806
36. Grandin CB, Duprez TP, Smith AM et al (2002) Which MR-derived perfusion parameters are the best predictors of infarct growth in hyperacute stroke? Comparative study between relative and quantitative measurements. Radiology 223:361-370
37. Rashid W, Parkes LM, Ingle GT et al (2004) Abnormalities of cerebral perfusion in multiple sclerosis. J Neurol Neurosurg Psychiatry 75:1288-1293

38. Sims DE (1991) Recent advances in pericyte biology: implications for health and disease. Can J Cardiol 7:431-443
39. Allan SM, Rothwell NJ (2003) Inflammation in central nervous system injury. Philos Trans R Soc Lond B Biol Sci 358:1669-1677
40. Broom KA, Anthony DC, Blamire AM et al (2005) MRI reveals that early changes in cerebral blood volume precede blood-brain barrier breakdown and overt pathology in MS-like lesions in rat brain. J Cereb Blood Flow Metab 25:204-216
41. Ge Y, Law M, Johnson G et al (2004) Basal ganglia neurodegeneration in multiple sclerosis measured by dynamic susceptibility contrast MRI. Proc Intl Soc Mag Reson Med, Kyoto, Japan
42. Lycke J, Wikkelso C, Bergh AC et al (1993) Regional cerebral blood flow in multiple sclerosis measured by single photon emission tomography with technetium-99m hexamethylpropyleneamine oxime. Eur Neurol 33:163-167
43. Ge Y, Gonen O, Inglese M et al (2004) Neuronal cell injury precedes brain atrophy in multiple sclerosis. Neurology 62:624-627
44. Roser W, Hagberg G, Mader I et al (1995) Proton MRS of gadolinium-enhancing MS plaques and metabolic changes in normal-appearing white matter. Magn Reson Med 33:811-817
45. Rovira A, Pericot I, Alonso J et al (2002) Serial diffusion-weighted MR imaging and proton MR spectroscopy of acute large demyelinating brain lesions: case report. Am J Neuroradiol 23:989-994
46. Dietrich RB, Bradley WG Jr (1988) Iron accumulation in the basal ganglia following severe ischemic-anoxic insults in children. Radiology 168:203-206
47. Bakshi R, Dmochowski J, Shaikh ZA, Jacobs L (2001) Gray matter T2 hypointensity is related to plaques and atrophy in the brains of multiple sclerosis patients. J Neurol Sci 185:19-26
48. Buettner GR, Schafer FQ (2000) Free radicals, oxidants, and antioxidants. Teratology 62:234
49. Levine SM, Chakrabarty A (2004) The role of iron in the pathogenesis of experimental allergic encephalomyelitis and multiple sclerosis. Ann N Y Acad Sci 1012:252-266

Chapter 7

Diffusion-Weighted Imaging

M. Rovaris, E. Perego, M. Filippi

Introduction

Conventional magnetic resonance imaging (MRI) is widely used for the diagnosis and monitoring of multiple sclerosis (MS), because it is more sensitive than clinical assessment in detecting disease dissemination over space and time [1] and for revealing the occurrence of disease activity and the accumulation of disease burden over time. Nevertheless, the discrepancies between clinical and conventional MRI findings in patients with established MS [2] highlight the fact that conventional MRI is unable to reliably assess the more disabling pathological features of the disease, including axonal and neuronal loss.

During the last decade, diffusion-weighted imaging (DWI) is increasingly being applied to the study of MS, because of its ability to detect and quantify disease-related changes in the tissue microstructure within and outside T2-visible lesions. Since diffusion can be defined as the microscopic random translational motion of molecules in a fluid system, in the central nervous system (CNS), this parameter is influenced by several tissue components, including cell membranes and organelles. The MRI-measured diffusion coefficient of healthy biological tissues is, therefore, lower than that in free water and is called the apparent diffusion coefficient (ADC) [3]. Pathological processes which result in a loss or increased permeability of "restricting" barriers can determine an increase in the ADC values. Since the magnitude of diffusion is dependent on the direction in which it is measured, DWI can also give information about the geometry of tissue structures [4]. A full characterization of diffusion can be obtained in terms of a tensor [5], a 3×3 matrix which accounts for the correlation existing between molecular displacement along orthogonal directions. From the tensor, it is possible to derive the mean diffusivity (MD), which is a measure of diffusivity independent of the spatial orientation of tissue structures, and some other dimensionless indices of anisotropic diffusion, including fractional anisotropy (FA) [6, 7], which reflect the prevalence of diffusivity along one spatial direction (e.g., along axonal fibers rather than perpendicular to them). Against this background, the pieces of evidence supporting a role for DWI in the study of neurodegeneration in MS and in monitoring its evolution over time will be reviewed critically.

DWI and Neurodegeneration in MS

Pathological Specificity

In principle, both demyelination and axonal degeneration have the potential to alter the permeability or geometry of structural barriers to water molecular diffusion in the brain, thus leading to DWI-detectable changes. A high-field ex vivo DWI study of post-mortem MS spinal cord (four cases) has revealed abnormalities of ADC and an index of anisotropy when compared with one healthy control sample [8]. The anisotropy measure provided a better correlation with axonal density (r=0.61, p<0.001) and myelin content (r=0.51, p<0.001) than did ADC (r=–0.32, p=0.017 and r=–0.45, p=0.001, respectively). Moreover, the results of DWI studies of aging [9] and Alzheimer's disease [10, 11] also support the notion that this technique has the potential to quantify the severity of degenerative changes in the human brain.

The pathological specificity of DWI for neurodegeneration is also highlighted by the available studies of T2-visible MS lesions [12-24] (Fig. 1). Despite always

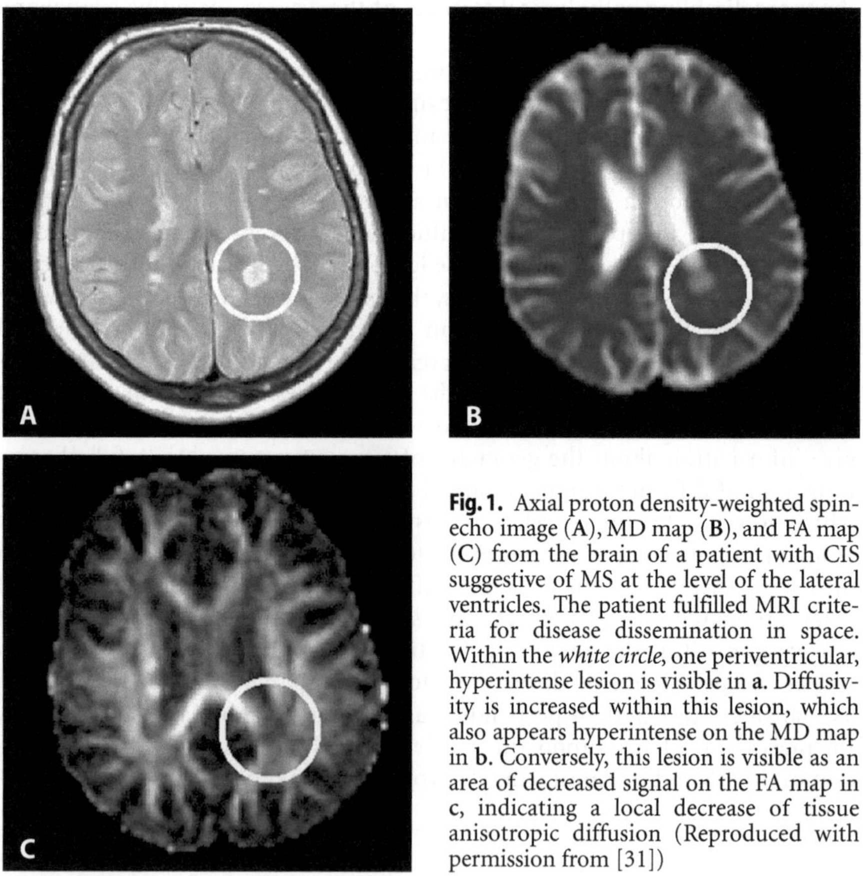

Fig. 1. Axial proton density-weighted spin-echo image (**A**), MD map (**B**), and FA map (**C**) from the brain of a patient with CIS suggestive of MS at the level of the lateral ventricles. The patient fulfilled MRI criteria for disease dissemination in space. Within the *white circle*, one periventricular, hyperintense lesion is visible in **a**. Diffusivity is increased within this lesion, which also appears hyperintense on the MD map in **b**. Conversely, this lesion is visible as an area of decreased signal on the FA map in **c**, indicating a local decrease of tissue anisotropic diffusion (Reproduced with permission from [31])

being more pronounced than those found in the normal-appearing brain tissues, water diffusion abnormalities do vary in different types of MS lesions. All investigators have shown higher ADC or MD and lower FA values in T1-hypointense than in T1-isointense lesions [13, 14, 16, 17, 21-24]. Since the former lesions represent areas of irreversible tissue disruption and axonal loss [25], this finding confirms that DWI may have a role in quantifying the extent of neurodegenerative changes associated with MS lesions. On the other hand, conflicting results have been obtained when comparing ADC or MD values in enhancing versus non-enhancing lesions [13, 14, 16, 17, 21, 23]. Several studies have shown that water diffusivity is markedly increased in ring-enhancing lesions when compared to homogeneously enhancing lesions [21-23]. Markedly reduced FA values have also been found in ring-enhancing lesions [21]. Since ring-enhancing areas almost always correspond to chronic, re-activated MS lesions, these findings also suggest that DWI is sensitive to the more destructive aspects of MS pathology. A DWI study of large MS lesions which were followed-up for 1-3 months [26] has shown that MD values were increased in all acute lesions at baseline, but they continued to increase during follow-up only in a subgroup of these lesions, which might represent those with more pronounced tissue damage. Additional longitudinal studies are needed in order to address the important issue of how much of this DWI-detectable tissue disorganization of enhancing lesions is permanent (i.e., related to neurodegeneration) and how much is transient (i.e., related to edema, demyelination, and remyelination).

Sensitivity

Numerous region-of-interest and histogram-based DWI studies of MS have consistently shown the presence of diffusion abnormalities in the normal-appearing white matter (NAWM) and gray matter (GM) of patients with MS (see [27] for a review), which can even precede the development of T2-visible lesions by several weeks [15, 20]. An exception to this is the study by Griffin et al. [28], who did not find any significant difference in MD and FA values of NAWM regions between patients with early relapsing/remitting (RR) MS and controls. However, a subsequent study of this cohort using segmented histogram analysis did report subtle abnormalities of FA in the normal-appearing brain tissue [29]. A preliminary study of patients at presentation with clinically isolated syndrome (CIS) suggestive of MS [30] also failed to detect significant differences of NAWM ADC between patients and healthy controls. However, Gallo et al. [31], using histogram-based analysis of segmented diffusion maps of the brain, reported a significant increase in MD and decrease in FA values in the NAWM of CIS patients with paraclinical evidence of disease dissemination in space, although the changes did not predict the short-term occurrence of conventional MRI disease activity. The results of other studies conducted in patients with progressive MS indicate that diffusion abnormalities become more pronounced with increasing disease duration and neurological impairment [32-34], thus supporting the notion that DWI is particularly sensitive to the more disabling features of MS pathology. The observed limited correlations between DWI findings, the corresponding lesion burden, and other MRI measures of tissue damage in the NAWM/GM suggest that diffusivity changes do not merely depend upon neuronal/axonal retrograde degeneration, but may reflect the presence of pathological features (e.g., microscopic lesions), which may go undetected when using conventional MRI. These findings also suggest that DWI may complement other techniques when assessing the actual burden of the disease in MS patients.

The sensitivity of DWI to the evolution of irreversible MS damage over short-term periods of time has been highlighted by several studies. A 1-year follow-up study of CIS patients [30] found that NAWM ADC values become significantly higher than those of healthy controls only on follow-up imaging and are significantly correlated with T2 lesion load at the same time-point. In another study of untreated patients with RRMS, a significant decrease in average GM FA and a significant increase in average GM MD were observed over 18 months of follow-up [35]. No correlation was found between changes in average GM MD, GM volume, and T2 lesion load. A significant worsening of GM DWI histogram-derived parameters after 1 year of follow-up has also been described in a large cohort of progressive MS patients [34], without any relationship with the concomitant lesion load accumulation. In contrast with this, however, Schmierer et al. [36] observed an increased diffusivity of NAWM regions, but not of the thalamus, in a small cohort of primary progressive (PP) MS patients who were followed up for 1 year.

Correlation with Clinical Findings

Significant correlations between DWI findings and MS clinical manifestations or disability were not found in several studies [12, 13, 16, 18, 28], perhaps because of the relatively small samples studied [12], the limited brain coverage [12-13], or the narrow range of disabilities that was considered [18, 28]. With improved DWI acquisition schemes and greater numbers of patients studied, correlations between DWI and clinical findings in MS are now emerging [17, 37-45]. Interestingly, the strongest clinical/DWI relationships were found for the diffusion characteristics of T2-visible lesions [17, 38] and GM [32, 42, 43]. In an exploratory study, Rovaris et al. [42] found a moderate correlation between a global cognitive impairment index and MD of the GM in mildly disabled RRMS patients. On the other hand, Ciccarelli et al. [38] found a strong correlation between FA in the supratentorial and infratentorial NAWM and neurological disability in patients with RRMS. Moreover, patients with different MS phenotypes showed a relationship between MD and FA changes in the cerebral peduncles and pyramidal functional scores [38]. In another study [44], a moderate correlation was found between anisotropy indexes from the corpus callosum and disability scores in patients with PPMS. These findings indicate that the pathologic damage detected by DWI in clinically eloquent brain regions is a significant factor contributing to disability in MS. Two preliminary cross-sectional studies [45, 46] also suggest that DWI may contribute to the generation of composite MR-based scores able to explain a great part of the variance in MS-related disability. Moreover, that DWI may be sensitive to the more disabling features of MS pathology is indicated by the results of a prospective study of patients with PPMS [47], for whom the severity of diffusion abnormalities in the GM was found to be a significant predictor of disability worsening 5 years later.

Optic Nerve and Spinal Cord Studies

DWI of the optic nerve and spinal cord presents more technical difficulties than that of the brain, and studies of these structures have therefore been limited. In a study of 16 patients, 1 year after an attack of optic neuritis, the mean ADC from diseased optic nerves was significantly higher than those from healthy contralateral (p=0.005) and control optic nerves (p=0.006) [48]. The ADC was strongly correlated with both visual acuity and visual evoked potential (VEP) parameters. A recent, histogram-based DWI study [49] of the cervical cord of 44 patients with RRMS and secondary progressive (SP) MS has shown that, compared with controls, MS patients had significantly higher average MD and MD histogram peak location and lower average FA and FA histogram peak location. SPMS patients also had significantly higher average MD and MD histogram peak location than controls and lower average FA than RRMS patients. A multivariate linear regression model retained cord average FA and brain average MD as variables independently influencing patients' clinical disability. The same acquisition and post-processing techniques were also used to study a sample of 24 PPMS patients [50], in whom a significant decrease in average FA and increase in average MD was found when compared with healthy subjects. All these findings indicate that DWI of the optic nerve and the spinal cord may allow us to study neurodegeneration in these clinically eloquent CNS regions.

Future Perspectives

Due to the different tissue anisotropies, DWI is also able to elucidate the macroscopic volume-averaged orientation of the microstructure therein. The examination of diffusion tensor (DT) eigenvectors [51], which represent the diffusion coefficients along the axes of the diffusion ellipsoid, may provide a signature of Wallerian degeneration in the NAWM of RRMS patients. This pathological feature is associated with an increased diffusion transverse to the fibers, but not along them. The anisotropy also provides the basis of DT tractography methods, developed to determine in vivo the pathways of anatomical CNS connections. Recently, several authors have developed different tractography techniques that utilize the directional and anisotropy information contained in a given voxel to connect neighboring voxels and to visualize brain white-matter tracts [52-54] (Fig. 2). The application of tractography methods to segment the different functional white-matter structures in MS patients might allow the strength of the correlations between clinical and MRI findings to be improved. Wilson et al. [55] produced maps of corticospinal tracts (CST) from DWI scans of 25 patients with RRMS and measured the relative anisotropy along these pathways. Significant correlations were found between the latter parameter and patients' EDSS and pyramidal functional system scores, whereas T2 lesion burden and diffusion histogram parameters did not correlate with clinical findings. Vaithianathar et al. [56] used DT-based tractography to map the CST of 25 patients with RRMS. They then sampled T1 relaxation time values along the corresponding trajectories on co-registered whole-brain T1 maps. CST T1, but not total white-matter T1 values, correlated significantly with the severity of patients' neurological disability. Pagani et al. [57] using an ad hoc post-processing technique for DT fiber tracking of the CST in CIS patients, found that a regional increase in transverse diffusion eigenvalues was correlated with the presence of motor impairment. In patients with RRMS, Lin et al. [58] mapped the CST and the corpus callosum using DT tractography and correlated the ADC values in these regions with clinical measures thought to specifically reflect their functioning. These measures were the patients' scores on the pyramidal functional system and paced auditory serial addition test (PASAT), the latter being an index of short-term memory and sustained attention. Correlation analysis showed that ADC of the CST explained about 25% of the pyramidal score variance, while ADC of the corpus callosum explained more than 33% of the PASAT score variance. The results of the latter studies [57, 58] support the notion that, thanks to the application of DT tractography, the strategy of "critical sampling" in MS may enable us to quantify in vivo the severity of neurodegenerative features related to specific clinical impairment.

Newer acquisition schemes for DWI, such as high b-value q-space images [59, 60], may further increase the sensitivity of "conventional" low b-value DWI in the detection of NAWM abnormalities. This might be particularly relevant for pathological features related to axonal dysfunction or degeneration, as suggested

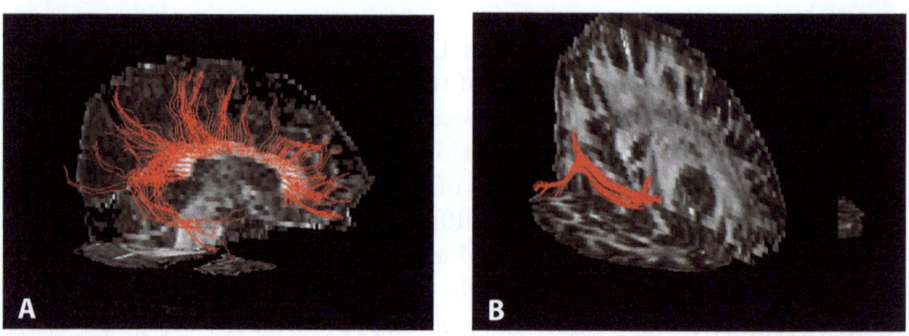

Fig. 2. 3D projections of DT tracking of the corpus callosum (A) and optic radiation (B) fibers in a right-handed, 28-year-old healthy woman

by the results of a study [61] showing that q-space diffusion displacement values were correlated with the levels of N-acetylaspartate (an MR spectroscopy marker of axonal viability) in the brain white matter of MS patients and healthy controls.

Conclusions

DWI is sensitive to the more destructive aspects of MS pathology, including neurodegeneration, and enables us to quantify their severity both within and outside T2-visible lesions. The preliminary results of longitudinal studies indicate that DWI may also be able to track the short-term changes in MS-related damage within lesions and in the GM. Thanks to the application of DT tractography and to the availability of acquisition schemes for studying the optic nerve and the spinal cord, DWI may well be able to characterize the patterns of MS neurodegeneration in clinically eloquent CNS regions. Such a strategy of "critical sampling" [62], which would also include the DWI study of GM damage, although it might contribute to overcoming the "clinical/MRI paradox" of MS, should nevertheless be viewed as a component of a more comprehensive approach to the disease work-up.

Little is still known about the actual features underlying diffusion changes in MS. On the other hand, however, several pieces of evidence suggest that the various (and often concomitant) pathological changes occurring in the MS brain might affect the diffusivity and anisotropy characteristics of tissues in opposite ways, thereby reducing the sensitivity and specificity of DWI findings. The best acquisition and post-processing strategies for MS studies remain a matter of debate, and the contribution of newer and more sophisticated techniques to DWI investigations in MS needs to be evaluated further. Finally, the stability and sensitivity to longitudinal, MS-related changes in DWI scans need to be tested and confirmed. All of this underpins the need for designing ad hoc protocols for diffusion imaging acquisition and post-processing in MS studies, which should be validated by postmortem investigations correlating pathological and diffusion findings in MS.

References

1. McDonald WI, Compston A, Edan G et al (2001) Recommended diagnostic criteria for multiple sclerosis: guidelines from the international panel on the diagnosis of multiple sclerosis. Ann Neurol 50:121-127
2. Filippi M, Grossman RI (2002) MRI techniques to monitor MS evolution: the present and the future. Neurology 58:1147-1153
3. Le Bihan D, Breton E, Lallemand D et al (1986) MR imaging of intravoxel incoherent motions: application to diffusion and perfusion in neurologic disorders. Radiology 161:401-407
4. Le Bihan D, Turner R, Pekar J, Moonen CTW (1991) Diffusion and perfusion imaging by gradient sensitization: design, strategy and significance. J Magn Reson Imaging 1:7-8
5. Basser PJ, Mattiello J, Le Bihan D (1994) Estimation of the effective self-diffusion tensor from the NMR spin-echo. J Magn Reson B 103:247-254

6. Pierpaoli C, Jezzard P, Basser PJ et al (1996) Diffusion tensor MR imaging of the human brain. Radiology 201:637-648
7. Basser PJ, Pierpaoli C (1996) Microstructural features measured using diffusion tensor imaging. J Magn Reson B 111:209-219
8. Mottershead JP, Schmierer K, Clemence M et al (2003) High field MRI correlates of myelin content and axonal density in multiple sclerosis: a post-mortem study of the spinal cord. J Neurol 250:1293-1301
9. Rovaris M, Iannucci G, Cercignani M et al (2003) Age-related changes in conventional, magnetization transfer, and diffusion-tensor MR imaging findings: study with whole-brain tissue histogram analysis. Radiology 227:731-738
10. Bozzali M, Franceschi M, Falini A et al (2001) Quantification of tissue damage in AD using diffusion tensor and magnetization transfer MRI. Neurology 57:1135-1137
11. Bozzali M, Falini A, Franceschi M et al (2002) White matter damage in Alzheimer's disease assessed in vivo using diffusion tensor magnetic resonance imaging. J Neurol Neurosurg Psychiat 72:742-746
12. Horsfield MA, Lai M, Webb SL et al (1996) Apparent diffusion coefficients in benign and secondary progressive multiple sclerosis by nuclear magnetic resonance. Magn Reson Med 36:393-400
13. Droogan AG, Clark CA, Werring DJ et al (1999) Comparison of multiple sclerosis clinical subgroups using navigated spin echo diffusion-weighted imaging. Magn Reson Imaging 17:653-661
14. Werring DJ, Clark CA, Barker GJ et al (1999) Diffusion tensor imaging of lesions and normal-appearing white matter in multiple sclerosis. Neurology 52:1626-1632
15. Werring DJ, Brassat D, Droogan AG et al (2000) The pathogenesis of lesions and normal-appearing white matter changes in multiple sclerosis: a serial diffusion MRI study. Brain 123:1667-1676
16. Filippi M, Iannucci G, Cercignani M et al (2000) A quantitative study of water diffusion in multiple sclerosis lesions and normal-appearing white matter using echo-planar imaging. Arch Neurol 57:1017-1021
17. Filippi M, Cercignani M, Inglese M et al (2001) Diffusion tensor magnetic resonance imaging in multiple sclerosis. Neurology 56:304-311
18. Cercignani M, Iannucci G, Rocca MA et al (2000) Pathologic damage in MS assessed by diffusion-weighted and magnetization transfer MRI. Neurology 54:1139-1144
19. Cercignani M, Bozzali M, Iannucci G et al (2001) Magnetisation transfer ratio and mean diffusivity of normal appearing white and grey matter from patients with multiple sclerosis. J Neurol Neurosurg Psychiat 70:311-317
20. Rocca MA, Cercignani M, Iannucci G et al (2000) Weekly diffusion-weighted imaging of normal-appearing white matter in MS. Neurology 55:882-884
21. Bammer R, Augustin M, Strasser-Fuchs S et al (2000) Magnetic resonance diffusion tensor imaging for characterizing diffuse and focal white matter abnormalities in multiple sclerosis. Magn Reson Med 44:583-589
22. Nusbaum AO, Lu D, Tang CY, Atlas SW (2000) Quantitative diffusion measurements in focal multiple sclerosis lesions: correlations with appearance on T1-weighted MR images. Am J Roentgenol 175:821-825
23. Roychowdhury S, Maldijan JA, Grossman RI (2000) Multiple sclerosis: comparison of trace apparent diffusion coefficients with MR enhancement pattern of lesions. Am J Neuroradiol 21:869-874
24. Castriota-Scanderbeg A, Fasano F, Hagberg G et al (2003) Coefficient D_{av} is more sensitive than fractional anisotropy in monitoring progression of irreversible tissue damage in focal nonactive multiple sclerosis lesions. Am J Neuroradiol 24:663-670
25. van Walderveen MA, Kamphorst W, Scheltens P et al (1998) Histopathologic correlate of hypointense lesions on T1-weighted spin-echo MRI in multiple sclerosis. Neurology 50:1282-1288

26. Castriota-Scanderbeg A, Sabatini U, Fasano F et al (2002) Diffusion of water in large demyelinating lesions: a follow-up study. Neuroradiology 44:764-767
27. Rovaris M, Gass A, Bammer R et al (2005) Diffusion MRI in multiple sclerosis. Neurology 65:1526-1532
28. Griffin CM, Chard DT, Ciccarelli O et al (2001) Diffusion tensor imaging in early relapsing-remitting multiple sclerosis. Mult Scler 7:290-297
29. Rashid W, Hadjiprocopis A, Griffin CM et al (2004) Diffusion tensor imaging of early relapsing-remitting multiple sclerosis with histogram analysis using automated segmentation and brain volume correction. Mult Scler 10:9-15
30. Caramia F, Pantano P, Di Legge S et al (2000) A longitudinal study of MR diffusion changes in normal appearing white matter of patients with early multiple sclerosis. Magn Reson Imaging 20:383-388
31. Gallo A, Rovaris M, Riva R et al (2005) Diffusion tensor MRI detects normal-appearing white matter damage unrelated to short-term disease activity in patients at the earlier stage of multiple sclerosis. Arch Neurol 62:803-808
32. Bozzali M, Cercignani M, Sormani MP, Comi G, Filippi M (2002) Quantification of brain gray matter damage in different MS phenotypes by use of diffusion tensor MR imaging. Am J Neuroradiol 23:985-988.
33. Rovaris M, Bozzali M, Iannucci G et al (2002) Assessment of normal-appearing white and gray matter in patients with primary progressive multiple sclerosis: a diffusion-tensor magnetic resonance imaging study. Arch Neurol 59:1406-1412
34. Rovaris M, Gallo A, Valsasina P et al (2005) Short-term accrual of gray matter pathology in patients with progressive multiple sclerosis: an in vivo study using diffusion tensor MRI. Neuroimage 24:1139-1146
35. Oreja-Guevara C, Rovaris M, Iannucci G et al (2005) Progressive gray matter damage in patients with relapsing-remitting MS: a longitudinal diffusion tensor MRI study. Arch Neurol 62:578-584
36. Schmierer K, Altmann DR, Kassim N et al (2004) Progressive change in primary progressive multiple sclerosis normal-appearing white matter: a serial diffusion magnetic resonance imaging study. Mult Scler 10:182-187
37. Cercignani M, Bozzali M, Iannucci G et al (2002) Intra-voxel and inter-voxel coherence in patients with multiple sclerosis assessed using diffusion tensor MRI. J Neurol 249:875-883
38. Ciccarelli O, Werring DJ, Wheeler-Kingshott CA et al (2001) Investigation of MS normal-appearing brain using diffusion tensor MRI with clinical correlations. Neurology 56:926-933
39. Cercignani M, Inglese M, Pagani E et al (2001) Mean diffusivity and fractional anisotropy histograms in patients with multiple sclerosis. Am J Neuroradiol 22:952-958
40. Nusbaum AO, Tang CY, Wei TC et al (2000) Whole-brain diffusion MR histograms differ between MS subtypes. Neurology 54:1421-1426
41. Castriota-Scanderbeg A, Tomaiuolo F, Sabatini U et al (2000) Demyelinating plaques in RR and secondary-progressive multiple sclerosis: assessment with diffusion MR imaging. Am J Neuroradiol 21:862-868
42. Rovaris M, Iannucci G, Falautano M et al (2002) Cognitive dysfunction in patients with mildly disabling relapsing-remitting multiple sclerosis: an exploratory study with diffusion tensor MR imaging. J Neurol Sci 195:103-109
43. Vrenken H, Pouwels PJ, Geurts JJ et al (2006) Altered diffusion tensor in multiple sclerosis normal-appearing brain tissue: cortical diffusion changes seem related to clinical deterioration. J Magn Reson Imaging 23:628-636
44. Oh J, Henry RG, Genain C et al (2004) Mechanisms of normal appearing corpus callosum injury related to pericallosal T1 lesions in multiple sclerosis using directional diffusion tensor and 1H MRS imaging. J Neurol Neurosurg Psychiat 75:1281-1286

45. Mainero C, De Stefano N, Iannucci G et al (2001) Correlates of MS disability assessed in vivo using aggregates of MR quantities. Neurology 56:1331-1334
46. Pulizzi A, Rovaris M, Judica E et al (2007) Determinants of disability in multiple sclerosis at various disease stages: a multiparametric magnetic resonance study. Arch Neurol (*in press*)
47. Rovaris M, Judica E, Gallo A et al (2006) Grey matter damage predicts the evolution of primary progressive multiple sclerosis at 5 years. Brain 129:2628-2634
48. Hickman SJ, Wheeler-Kingshott CAM, Jones SJ et al (2005) Optic nerve diffusion measurement from diffusion weighted imaging in optic neuritis. Am J Neuroradiol 26:951-956
49. Valsasina P, Rocca MA, Agosta F et al (2005) Mean diffusivity and fractional anisotropy histogram analysis of the cervical cord in MS patients. NeuroImage 26:822-828
50. Agosta F, Benedetti B, Rocca MA et al (2005) Quantification of cervical cord pathology in primary progressive MS using diffusion tensor MRI. Neurology 64:631-635
51. Henry RG, Oh J, Nelson SJ, Pelletier D (2003) Directional diffusion in relapsing-remitting multiple sclerosis: a possible in vivo signature of Wallerian degeneration. J Magn Reson Imaging 18:420-426
52. Conturo TE, Lori NF, Cull TS et al (1999) Tracking neuronal fiber pathways in the living human brain. Proc Natl Acad Sci USA 96:10422-10427
53. Mori S, Crain BJ, Chacko VP, van Zijl PC (1999) Three-dimensional tracking of axonal projections in the brain by magnetic resonance imaging. Ann Neurol 45:265-269
54. Mori S, Kaufmann WE, Davatzikos C et al (2002) Imaging cortical association tracts in the human brain using diffusion-tensor-based axonal tracking. Magn Reson Med 47:215-223
55. Wilson M, Tench CR, Morgan PS, Blumhardt LD (2003) Pyramidal tract mapping by diffusion tensor magnetic resonance imaging in multiple sclerosis: improving correlations with disability. J Neurol Neurosurg Psychiat 74: 203-207
56. Vaithianathar L, Tench CR, Morgan PS et al (2002) T1 relaxation time mapping of white matter tracts in multiple sclerosis defined by diffusion tensor imaging. J Neurol 249:1272-1278
57. Pagani E, Filippi M, Rocca MA, Horsfield MA (2005) A method for obtaining tract-specific diffusion tensor MRI measurements in the presence of disease: application to patients with clinically isolated syndromes suggestive of multiple sclerosis. NeuroImage 26:258-265
58. Lin X, Tench CR, Morgan PS et al (2005) 'Importance sampling' in MS: use of diffusion tensor tractography to quantify pathology related to specific impairment. J Neurol Sci 237:13-19
59. Assaf Y, Ben-Bashat D, Chapman J et al (2002) High b-value q-space analyzed diffusion-weighted MRI: application to multiple sclerosis. Magn Reson Med 47:115-126
60. Assaf Y, Cohen Y (2000) Assignment of the water slow-diffusing component in the central nervous system using q-space diffusion MRS: implications for fiber tract imaging. Magn Reson Med 43:191-199
61. Assaf Y, Chapman J, Ben-Bashat D (2005) White matter changes in multiple sclerosis: correlation of q-space diffusion MRI and 1H MRS. Magn Reson Imaging 23:703-710
62. Rovaris M, Filippi M (2005) "Importance sampling": a strategy to overcome the clinical/MRI paradox in MS? J Neurol Sci 237:1-3

Chapter 8

Proton MR Spectroscopy

N. De Stefano

Introduction

In the last few years, there has been an increased appreciation of the apparently primary role of neuronal and axonal injury in the pathogenesis of multiple sclerosis (MS) [1, 2]. This has been driven to a significant degree by the results of proton magnetic resonance (MR) spectroscopy (^{1}H-MRS) studies, which have emphasized that substantial neuroaxonal damage occurs inside the demyelinating lesions as well as in the normal-appearing white matter (NAWM) and gray matter (GM) of the brain of patients with MS [2, 3]. All this has been confirmed pathologically [1, 4, 5] and has led to a reconsideration of the role of axonal damage in MS [6].

Despite the specificity of ^{1}H-MRS and the relatively large number of clinical ^{1}H-MRS studies on patients with MS, measures provided by this technique are rarely used as MR endpoints in trials with disease-modifying agents. By contrast, findings of recent in-vivo and ex-vivo studies have shed new light into the pathogenesis of MS [7-9] and have pinpointed the need for pathologically specific surrogate markers that can monitor disease changes beyond focal demyelination.

Basics of Proton MR Spectroscopy

^{1}H-MRS is fundamentally similar to magnetic resonance imaging (MRI). However, while conventional MRI provides structural information based on signal from water, proton ^{1}H-MRS is adapted to record signals from metabolites present in tissues. Resonances in MR spectra are identified primarily by their frequency (i.e., position in the spectrum) and expressed as the shift in frequency in parts per million (ppm) relative to a standard. For spectroscopic data to be clinically useful, it is important to localize the source of the signals in order both to collect the signal exclusively from a given brain region and to eliminate unwanted signal from the fat that is present in the scalp and skull. The most frequently used ^{1}H-MRS localization techniques can be divided into *single-voxel* and *multivoxel types*. In the single-voxel localization, the signals are acquired from a "brick-shaped" single volume of various sizes (minimum 2-3 cc). In contrast, the multivoxel approach generates individual spectra from multiple volume elements (voxels, minimum volume 0.5-1 cc) at the same time. Most commonly, multivoxel ^{1}H-MRS is done in 2-D for one or more slices in the so-called MR spectroscopic imaging (^{1}H-MRSI). A third, newer, approach, which is reminiscent of one of the earliest in-vivo localization techniques (the topical MR [10]), relies on the homogeneity of the static magnetic field to collect the signal of a given metabolite from those parts of the brain, brainstem, and spinal cord that are within a sufficiently homogeneous static magnetic field and within the sensitive volume of the coil [11].

As water may be present in brain tissues at a thousand times greater concentration than any other metabolite and produces signals a thousand times greater than metabolite signals, it is common in ^{1}H-MRS studies to use special sequences to suppress the excessively large signals from water. Water-suppressed localized ^{1}H-MRS of the brain can be performed at long or short echo times. At long echo times, the human brain reveals four major resonances [12]: (i) at 3.2 ppm, arising mainly from tetramethylamines, especially choline-containing phospholipids (Cho); (ii) at 3.0 ppm, arising primarily from creatine (Cr), either alone or as phosphocreatine; (iii) at 2.0 ppm, arising from N-acetyl groups, especially N-acetylaspartate (NAA); and (iv) a doublet at 1.3 ppm, arising from the methyl resonance of lactate (Lac), which is normally barely visible above the baseline noise. Long echo time ^{1}H-MRS tends to be less influenced by gradient-induced distortions and provides a better defined baseline than short echo time measurements. Despite this, due to their relatively short T2 relaxation times, signals from lipids, myo-inositol (mI) and macromolecules are preferably obtained with short echo time measurements.

The quantification of individual metabolite concentrations from MR spectra obtained in vivo is challenging. One approach is simply to normalize signal intensities to an intravoxel standard such as Cr, in order to obtain an estimate of relative concentration. The main limitation of this approach is that Cr may not remain unaffected by MS pathology. Absolute quantification of ^{1}H-MRS metabolites can be achieved by quantifying the metabolites in the spectra with reference to a known internal standard (e.g., water) or to a known external standard (e.g., a phantom that contains a known concentration of a molecule of interest). Metabolite concentrations can then be expressed in arbitrary units or molar units. Unfortunately, tissue water does not provide a reliable reference, as it may vary between individuals (e.g., with nutritional or hydration state) and in response to pathology. The most common method for absolute quantification, therefore, uses an exogenous (phantom) reference [13].

Metabolic Changes in MS Brains as Detected by ^{1}H-MRS

In demyelinating lesions, ^{1}H-MRS at both short and long echo times reveals increases in Cho and Lac resonance intensities from the early phases of the pathological process [14, 15]. Changes in the resonance intensity of Cho result mainly from increases in the steady state levels of membrane phospholipids that are released during active myelin breakdown. Increases in Lac are likely to reflect the metabolism of inflammatory cells. In large, acute, demyelinating lesions decreases in Cr can also be seen [15]. Short echo time spectra give evidence for transient increases in visible lipids (released during myelin breakdown) and mI [16]. All these changes consistently include substantial decreases in NAA [2] and, since NAA is a metabolite detected almost exclusively in neurons and their processes in normal mature brains [15], the decreases in NAA have been interpreted as being due to axonal injury. Recently, glutamate levels were found to be elevated in acute lesions, suggesting a link between axonal injury in active lesions and impaired glutamate metabolism [17].

After the acute phase and over a period of days to weeks, there is a progressive reduction in raised Lac resonance intensities to normal levels in the lesional tissue [15]. Resonance intensities of Cr also seem to return to normal within a few days [15] or may show small increases, probably due to the presence of gliosis. This may also cause the persistence of increased mI resonance intensity in chronic lesions [18]. Resonance intensities of Cho and lipids seem to return to normal over a period of months [15, 16]. The signal intensity of NAA may remain decreased or show partial recovery, starting soon after the acute phase and lasting for several months [19]. The recovery of NAA may be related to resolution of edema, increases in the diameter of previously shrunk axons secondary to remyelination, and reversible metabolic changes in neurons [19].

Recent studies exploiting the greater coverage and resolution of ^{1}H-MRSI have shown that metabolic abnormalities in MS patients are not restricted to lesions [16, 20-24]. Thus, large NAA decreases were found in the NAWM of MS patients and interpreted as being due to diffuse axonal damage [25]. The extent of this NAA reduction decreases with the distance from the core of a lesion [26], consistent with the notion that the diffuse changes are, at least in part, related to dying-back of axons transected within plaques [1]. However, decreased levels of NAA seem to occur in the cerebral NAWM even without any obvious relation with T2-visible lesions [27-29].

In addition to NAA abnormalities, recent ^{1}H-MRS studies of patients both with clinically isolated syndrome suggestive of MS [30] and with established MS [18, 31] have demonstrated large increases in mI well outside the T2-hyperintense lesions, suggesting a significant increase in glial cell activity in apparently normal white matter of MS patients. Moreover, other ^{1}H-MRS data, by showing signals from lipids [16] or Cho [32] in regions that only later will develop new T2-hyperintense lesions, add to other quantitative MR data [33] in suggesting that focal myelin pathology may antedate the development of acute, severe inflammation. In this brain region, the intense gliosis may sometimes cause increases in Cr [34]. Glutamate may also be found to be increased [17].

As more sensitive MR techniques have been used, the contribution of GM to the total brain pathology in MS has also been shown to be substantial [35]. In this context, ^{1}H-MRS studies have shown significant decreases of NAA in the neocortical and subcortical GM of MS patients, consistent with neuronal or axonal injury [36-38]. In some studies [36, 39], ^{1}H-MRS and histopathologic methods have been used in combination, and the amount of ex-vivo total loss of thalamic neurons was largely comparable to the extent of neurodegeneration measured in vivo by the decrease in relative NAA concentration.

^{1}H-MRS to Assess Neurodegeneration in MS

Results of clinical ^{1}H-MRS studies have consistently reported decreases of NAA in lesions and/or in normal-appearing tissues in patients with the different forms of MS [35]. Brain NAA is synthesized in neuronal mitochondria from L-aspartate and acetyl-CoA and, although a number of possible functions have been suggested for NAA [40, 41], its physiological role is still not known. The fact that NAA is present almost exclusively in neurons of the mature human brain [42, 43] makes it a powerful surrogate marker of neuronal integrity in the brain, and decreases in NAA have been widely used as an indicator of brain pathology and disease progression in a variety of central nervous system diseases, including MS [35].

Most of the aforementioned clinical ^{1}H-MRS studies, however, report cerebral NAA normalized for intravoxel Cr (collected from large brain volumes located so as to obtain signal mainly from WM). This implies that changes of Cr may be, at least in part, responsible for the observed NAA/Cr changes [44-46]. As Cr is present in both neurons and glial cells [47], Cr changes can be indicative of neuroaxonal alteration (which is probably associated with decreases in both NAA and Cr signals), oligodendroglial disturbance (which is probably associated with decreases in Cr signals), and astrocytic proliferation (which is probably associated with increases in Cr signals) [34]. Thus, decreases in brain NAA/Cr could be interpreted as an indicator of an overall disturbance in the "cerebral tissue integrity" of that particular brain region [34]. In agreement with this, a number of spectroscopic studies have demonstrated highly significant correlations between NAA/Cr and clinical disability in patients with isolated acute demyelinating lesions and in patients with established MS [see, e.g., 15, 48, 49]. In general, this correlation is far from perfect and is often reported in relapsing/remitting (RR) MS patients or in patients at early disease stages, but is somewhat less close in patients with the progressive phase of MS [50].

In addition to NAA and Cr, other metabolites seem to have clinical importance. As well as the previously mentioned importance of Cho and lipid signals as predictors of the development of acute, severe inflammation [16, 32], diffuse increases of mI reported in the brain tissues of patients with different forms of the disease suggest that an intense glial proliferation process occurs in the MS brain [51] and can have relevance to clinical disability [30, 31, 52]. In contrast, increases of glutamate in lesional and non-lesional brain tissue seem to be related, at least in part, to the ongoing neurodegeneration, based on the hypothesis that an excess of glutamate in active lesions can be an important predictor of axonal injury, brain atrophy, and long-term accumulation of clinical disability [17, 53].

^{1}H-MRS to Monitor Neurodegeneration in MS

The above results strongly suggest that ^{1}H-MRS has the potential to monitor the temporal evolution of metabolite changes in both demyelinating lesions [15] and normal-appearing brain tissues [23, 26]. It is true, however, that longitudinal studies exploiting these unique ^{1}H-MRS characteristics are, in general, very rare.

Most of the longitudinal studies have concentrated on monitoring NAA or NAA/Cr changes (see [2] for a review), as the earliest studies of patients with demyelinating disease have emphasized that a substantial proportion of the decreases in NAA are transient in the acute phase of demyelination [14, 19]. Thus, changes over time of NAA or NAA/Cr levels have been shown to be closely correlated with changes in clinical disability in both patients with isolated acute demyelinating lesions [15] and in those with established MS followed through periods of relapse and remission [54]. Interestingly, this was also true in a single relapsing patient who was followed up for 6 years [55]. Consistent with other evidence of widespread pathology in MS, a strong correlation has also been found between changes in clinical disability and longitudinal changes in NAA/Cr in NAWM [23]. The latter, however, was not found in a more recent study of 20 patients with RRMS, which showed no correlations between NAA values and clinical scores over a 2-year follow-up period [56]. Overall these data provide evidence for the use of NAA or NAA/Cr as specific measures of pathological changes. Although these are probably less sensitive to changes with time than other MR indices (i.e., T2-weighted lesion volumes), they nevertheless seem appropriate for monitoring the progression of disability [57].

^{1}H-MRS has been used in a few small longitudinal studies to test whether disease-modifying therapies can reverse or arrest the progression of brain tissue injury [58-62]. These studies have reported conflicting results regarding the beneficial effects of the drug therapy, probably also due to the small cohorts of patients studied.

Despite this, however, there are rare examples of multicenter studies [27, 63] that, by reporting comparable cross-sectional ^{1}H-MRS values in normal controls at different centers, suggest that differences between sites can be minimized by standardizing the acquisition protocol and the post-processing procedures. In agreement with this, the reproducibility over time of ^{1}H-MRS measurements in healthy subjects suggests that metrics derived from ^{1}H-MRS could be used in treatment trials [64].

Thus, a standardized ^{1}H-MRS protocol has been used in some of the most recent, large, phase-III clinical trials, where a small proportion of the MR participating centers arranged the acquisition of a uniform single-voxel ^{1}H-MRS that was then analyzed in a centralized fashion. Preliminary results of one of these trials [65], which showed no changes over time in NAA/Cr after ineffective therapy, confirm that the use of brain metabolite levels as an index of disease outcome in clinical trials is feasible.

Conclusions

Evidence from ^{1}H-MRS studies indicates that the progression of axonal injury in MS is diffuse, inexorable, and accompanies even the earliest inflammatory changes. These results, confirmed by modern histopathology [66], have led to a reconsideration of the importance of neuronal and axonal pathology in this disease.

Despite its greater pathological specificity for axonal integrity compared with conventional MRI, the advantages of using outcome measures derived from ^{1}H-MRS in MS clinical trials have yet to be proven. Future studies and the few clinical trials that are currently incorporating ^{1}H-MRS into their imaging protocols will reveal whether ^{1}H-MRS has a role in quantifying the impact of therapeutic intervention on tissue damage in MS.

References

1. Trapp BD, Peterson J, Ransohoff RM et al (1998) Axonal transection in the lesions of multiple sclerosis. N Engl J Med 338:278-285
2. Matthews PM, De Stefano N, Narayanan S et al (1998) Putting magnetic resonance spectroscopy studies in context: axonal damage and disability in multiple sclerosis. Semin Neurol 18:327-336
3. Wolinsky JS, Narayana PA (2002) Magnetic resonance spectroscopy in multiple sclerosis: window into the diseased brain. Curr Opin Neurol 15:247-251
4. Ferguson B, Matyszak MK, Esiri MM, Perry VH (1997) Axonal damage in acute multiple sclerosis lesions. Brain 120:393-399
5. Peterson JW, Bo L, Mork S et al (2001) Transected neurites, apoptotic neurons, and reduced inflammation in cortical multiple sclerosis lesions. Ann Neurol 50:389-400
6. Trapp BD, Ransohoff RM, Fisher E, Rudick RA (1999) Neurodegeneration in multiple sclerosis: relationship to neurological disability. Neuroscientist 5:48-57
7. Waxman SG (2000) Multiple sclerosis as a neuronal disease. Arch Neurol 57:22-24
8. Trapp BD (2004) Pathogenesis of multiple sclerosis: the eyes only see what the mind is prepared to comprehend. Ann Neurol 55:455-457
9. Stys PK (2004) Axonal degeneration in multiple sclerosis: is it time for neuroprotective strategies? Ann Neurol 55:601-603
10. Gordon RE, Hanley PE, Shaw D et al (1980) Localization of metabolites in animals using 31P topical magnetic resonance. Nature 287:736-738
11. Gonen O, Viswanathan AK, Catalaa I et al (1998) Total brain N-acetylaspartate concentration in normal, age-grouped females: quantitation with non-echo proton NMR spectroscopy. Magn Reson Med 40:684-689
12. Arnold DL, Matthews PM (1996) Practical aspects of clinical applications of MRS in the brain. In: Young IR, Charles HC (eds) MR spectroscopy: clinical applications and techniques. Martin Dunitz, London, pp 139-159
13. Provencher SW (2001) Automatic quantitation of localized in vivo ^{1}H spectra with LCModel. NMR Biomed 14:260-264
14. Davie CA, Hawkins CP, Barker GJ et al (1994) Serial proton magnetic resonance spectroscopy in acute multiple sclerosis lesions. Brain 117:49-58
15. De Stefano N, Matthews PM, Antel JP et al (1995) Chemical pathology of acute demyelinating lesions and its correlation with disability. Ann Neurol 38:901-909
16. Narayana PA, Doyle TJ, Lai D, Wolinsky JS (1998) Serial proton magnetic resonance spectroscopic imaging, contrast-enhanced magnetic resonance imaging, and quantitative lesion volumetry in multiple sclerosis. Ann Neurol 43:56-71

17. Srinivasan R, Sailasuta N, Hurd R et al (2005) Evidence of elevated glutamate in multiple sclerosis using magnetic resonance spectroscopy at 3 T. Brain 128:1016-1025
18. Kapeller P, Brex PA, Chard D et al (2002) Quantitative ^{1}H MRS imaging 14 years after presenting with a clinically isolated syndrome suggestive of multiple sclerosis. Mult Scler 8:207-210
19. De Stefano N, Matthews PM, Arnold DL (1995) Reversible decreases in N-acetylaspartate after acute brain injury. Magn Reson Med 34:721-727
20. Husted CA, Goodin DS, Hugg JW et al (1994) Biochemical alterations in multiple sclerosis lesions and normal-appearing white matter detected by in vivo 31P and ^{1}H spectroscopic imaging. Ann Neurol 36:157-165
21. Narayanan S, Fu L, Pioro E et al (1997) Imaging of axonal damage in multiple sclerosis: spatial distribution of magnetic resonance imaging lesions. Ann Neurol 41:385-391
22. Davie CA, Barker GJ, Thompson AJ et al (1997) ^{1}H magnetic resonance spectroscopy of chronic cerebral white matter lesions and normal-appearing white matter in multiple sclerosis. J Neurol Neurosurg Psychiat 63:736-742
23. Fu L, Matthews PM, De Stefano N et al (1998) Imaging axonal damage of normal-appearing white matter in multiple sclerosis. Brain 121:103-113
24. Sarchielli P, Presciutti O, Pelliccioli GP et al (1999) Absolute quantification of brain metabolites by proton magnetic resonance spectroscopy in normal-appearing white matter of multiple sclerosis patients. Brain 122:513-521
25. Miller DH, Austin SJ, Connelly A et al (1991) Proton magnetic resonance spectroscopy of an acute and chronic lesion in multiple sclerosis. Lancet 337:58-59
26. De Stefano N, Narayanan S, Matthews PM et al (1999) In vivo evidence for axonal dysfunction remote from focal cerebral demyelination of the type seen in multiple sclerosis. Brain 122:1933-1939
27. De Stefano N, Narayanan S, Francis SJ et al (2002) Diffuse axonal and tissue injury in patients with multiple sclerosis with low cerebral lesion load and no disability. Arch Neurol 59:1565-1571
28. Wolinsky JS, Narayana PA, Fenstermacher MJ (1990) Proton magnetic resonance spectroscopy in multiple sclerosis. Neurology 40:1764-1769
29. Bruhn H, Frahm J, Merboldt KD et al (1992) Multiple sclerosis in children: cerebral metabolic alterations monitored by localized proton magnetic resonance spectroscopy in vivo. Ann Neurol 32:140-150
30. Fernando KT, McLean MA, Chard DT et al (2004) Elevated white matter myo-inositol in clinically isolated syndromes suggestive of multiple sclerosis. Brain 127:1361-1369
31. Chard DT, Griffin CM, McLean MA et al (2002) Brain metabolite changes in cortical grey and normal-appearing white matter in clinically early relapsing-remitting multiple sclerosis. Brain 125:2342-2352
32. Tartaglia MC, Narayanan S, De Stefano N et al (2002) Choline is increased in prelesional normal-appearing white matter in multiple sclerosis. J Neurol 249:1382-1390
33. Filippi M, Rocca MA, Martino G et al (1998) Magnetization transfer changes in the normal-appearing white matter precede the appearance of enhancing lesions in patients with multiple sclerosis. Ann Neurol 43:809-814
34. Caramanos Z, Narayanan S, Arnold DL (2005) ^{1}H-MRS quantification of tNA and tCr in patients with multiple sclerosis: a meta-analytic review. Brain 128:2483-2506
35. De Stefano N, Bartolozzi ML, Guidi L et al (2005) Magnetic resonance spectroscopy as a measure of brain damage in multiple sclerosis. J Neurol Sci 233:203-208
36. Wylezinska M, Cifelli A, Jezzard P et al (2003) Thalamic neurodegeneration in relapsing-remitting multiple sclerosis. Neurology 60:1949-1954
37. Inglese M, Liu S, Babb JS et al (2004) Three-dimensional proton spectroscopy of deep gray matter nuclei in relapsing-remitting MS. Neurology 63:170-172
38. Geurts JJ, Reuling IE, Vrenken H et al (2006) MR spectroscopic evidence for thalamic and hippocampal, but not cortical, damage in multiple sclerosis. Magn Reson Med 55:478-483
39. Cifelli A, Arridge M, Jezzard P et al (2002) Thalamic neurodegeneration in multiple sclerosis. Ann Neurol 52:650-653

40. Birken DL, Oldendorf WH (1989) N-acetyl-L-aspartic acid: a literature review of a compound prominent in ^{1}H-NMR spectroscopic studies of brain. Neurosci Biobehav Rev 13:23-31
41. Baslow MH (2003) N-acetylaspartate in the vertebrate brain: metabolism and function. Neurochem Res 28:941-953
42. Moffett JR, Namboodiri MAA, Cangro CB, Neale JH (1991) Immunohistochemical localization of N-acetylaspartate in rat brain. NeuroReport 2:131-134
43. Simmons ML, Frondoza CG, Coyle JT (1991) Immunocytochemical localization of N-acetyl-aspartate with monoclonal antibodies. Neuroscience 45:37-45
44. Vrenken H, Barkhof F, Uitdehaag BM et al (2005) MR spectroscopic evidence for glial increase but not for neuro-axonal damage in MS normal-appearing white matter. Magn Reson Med 53:256-266
45. Helms G, Stawiarz L, Kivisakk P, Link H (2000) Regression analysis of metabolite concentrations estimated from localized proton MR spectra of active and chronic multiple sclerosis lesions. Magn Reson Med 43:102-110
46. Filippi M, Falini A, Arnold DL et al (2005) Magnetic resonance techniques for the in vivo assessment of multiple sclerosis pathology: consensus report of the White Matter Study Group. J Magn Reson Imaging 21:669-675
47. Urenjak J, Williams SR, Gadian DG, Noble M (1993) Proton nuclear magnetic resonance spectroscopy unambiguously identifies different neural cell types. J Neurosci 13:981-989
48. Davie CA, Barker GJ, Webb S et al (1995) Persistent functional deficit in multiple sclerosis and autosomal dominant cerebellar ataxia is associated with axon loss. Brain 118:1583-1592
49. De Stefano N, Narayanan S, Francis GS et al (2001) Evidence of axonal damage in the early stages of multiple sclerosis and its relevance to disability. Arch Neurol 58:65-70
50. Matthews PM, De Stefano N, Narayanan S et al (1998) Putting magnetic resonance spectroscopy studies in context: axonal damage and disability in multiple sclerosis. Semin Neurol 18:327-336
51. Bitsch A, Bruhn H, Vougioukas V et al (1999) Inflammatory CNS demyelination: histopathologic correlation with in vivo quantitative proton MR spectroscopy. Am J Neuroradiol 20:1619-1627
52. Sastre-Garriga J, Ingle GT, Chard DT et al (2005) Metabolite changes in normal-appearing gray and white matter are linked with disability in early primary progressive multiple sclerosis. Arch Neurol 62:569-573
53. Werner P, Pitt D, Raine CS (2001) Multiple sclerosis: altered glutamate homeostasis in lesions correlates with oligodendrocyte and axonal damage. Ann Neurol 50:169-180
54. De Stefano N, Matthews PM, Fu L et al (1998) Axonal damage correlates with disability in patients with relapsing-remitting multiple sclerosis. Results of a longitudinal magnetic resonance spectroscopy study. Brain 121:1469-1477
55. De Stefano N, Matthews PM, Narayanan S et al (1997) Axonal dysfunction and disability in a relapse of multiple sclerosis: longitudinal study of a patient. Neurology 49:1138-1141
56. Tiberio M, Chard DT, Altmann DR et al (2006) Metabolite changes in early relapsing-remitting multiple sclerosis: a two year follow-up study. J Neurol 253:224-230
57. Mainero C, De Stefano N, Iannucci G et al (2001) Correlates of MS disability assessed in vivo using aggregates of MR quantities. Neurology 56:1331-1334
58. Sarchielli P, Presciutti O, Tarducci R et al (1998) ^{1}H-MRS in patients with multiple sclerosis undergoing treatment with interferon beta-1a: results of a preliminary study. J Neurol Neurosurg Psychiat 64:204-212
59. Narayanan S, De Stefano N, Francis GS et al (2001) Axonal metabolic recovery in multiple sclerosis patients treated with interferon beta-1b. J Neurol 248:979-986
60. Schubert F, Seifert F, Elster C et al (2002) Serial ^{1}H-MRS in relapsing-remitting multiple sclerosis: effects of interferon-beta therapy on absolute metabolite concentrations. MAGMA 14:213-222
61. Parry A, Corkill R, Blamire AM et al (2003) Beta-interferon treatment does not always slow the progression of axonal injury in multiple sclerosis. J Neurol 250:171-178

62. Khan O, Shen Y, Caon C et al (2005) Axonal metabolic recovery and potential neuroprotective effect of glatiramer acetate in relapsing-remitting multiple sclerosis. Mult Scler 11:646-651
63. Narayana PA, Wolinsky JS, Rao SB et al (2004) Multicentre proton magnetic resonance spectroscopy imaging of primary progressive multiple sclerosis. Mult Scler 10(Suppl 1):S73-S78
64. Leary SM, Brex PA, MacManus DG et al (2000) A ^{1}H magnetic resonance spectroscopy study of aging in parietal white matter: implications for trials in multiple sclerosis. Magn Reson Imaging 18:455-459
65. Narayanan S, De Stefano N, Pouwels PJ et al (2005) The effect of oral glatiramer acetate treatment on axonal integrity in multiple sclerosis: results from the multicentre CORAL MRS sub-study. Mult Scler 11(Suppl 1):S60
66. Trapp BD, Ransohoff R, Rudick R (1999) Axonal pathology in multiple sclerosis: relationship to neurologic disability. Curr Opin Neurol 12:295-302

Chapter 9

Functional MRI

M.A. Rocca, M. Filippi

Introduction

Over the past decade, modern structural magnetic resonance imaging (MRI) techniques have been extensively used for the study of patients with multiple sclerosis (MS) [1] with the ultimate goal of increasing our understanding of the mechanisms responsible for the accumulation of irreversible disability. The application of these techniques has provided important insights into the pathobiology of MS. First, it has been demonstrated that MS-related damage is not restricted to T2-visible lesions, but also involves, diffusely, the normal-appearing white matter (NAWM) and gray matter (GM) [1]. Secondly, it has been shown that the neurodegenerative component of the disease is not a late phenomenon and that it is not completely driven by inflammatory demyelination [2]. Finally, the contribution of axon damage to the clinical manifestations of the disease and to its clinical worsening over time has been confirmed [3, 4]. Despite these improvements, the correlation between the results of MRI and clinical findings remains suboptimal. This might be explained, at least partially, by the variable effectiveness of reparative and recovery mechanisms following MS-related tissue damage.

Although resolution of acute inflammation, remyelination, and redistribution of voltage-gated sodium channels in persistently demyelinated axons are all likely to limit the clinical impacts of damaging MS pathology [5], cortical reorganization has recently been suggested as an additional potential contributor to the recovery or to the maintenance of function in the presence of irreversible MS-related tissue damage. Brain plasticity is indeed a well-known feature of the human brain, which is likely to have several different substrates (including increased axonal expression of sodium channels, synaptic changes, increased recruitment of parallel existing pathways or "latent" connections, and reorganization of distant sites), and which may have a major adaptive role in limiting the functional consequences of axonal loss in MS [6].

This chapter provides an update of the current "state-of-the-art" of the application of functional MRI (fMRI) to the study of MS pathophysiology.

Technical Aspects

fMRI is a relatively new technique that is non-invasive, well tolerated, and widely used to study the neuronal mechanisms of central nervous system (CNS) functioning, and to define abnormal patterns of brain activation resulting from disease.

The signal changes seen during fMRI studies depend on the blood oxygenation level-dependent (BOLD) mechanism, which, in turn, involves changes in the transverse magnetization relaxation time (either T2* in a gradient-echo sequence, or T2 in a spin-echo sequence). These changes are attributable to differences in deoxyhemoglobin concentration subsequent to variations in neuronal activity [7]. The correlation between local deoxyhemoglobin levels and neuronal activity is thought to result from changes in oxygen extraction, cerebral blood flow (CBF), and cerebral blood volume (CBV) [7], all of which change with neuronal activity [8-10]. Activation of a cerebral tissue determines an increase in local synaptic activity, which results in a rise in blood flow and oxygen consumption. The increase in blood flow is greater than that in oxygen consumption, thus producing an increase in the ratio between oxygenated and deoxygenated hemoglobin, which enhances the MRI signal [10]. Because these signal changes are very small (usually ranging from 0.5% to 1.5%), a large number of brain images need to be acquired during alternating periods of activation (i.e., motor, sensorial, and cognitive) and rest [10]. By analyzing these data with the appropriate statistical methods, it is possible to obtain information about the location and extent of specific areas involved in the performance of a given task in healthy subjects and in patients [11, 12].

The main problem in the interpretation of fMRI studies is that the observed changes might be biased by differences in task performance between patients and controls. Clearly, this is a major issue in MS, which typically causes impairment of various functional systems. Therefore, despite providing several important pieces of information, the value of the earliest fMRI studies of patients with MS [13-19] has to be considered against this background.

For this reason, the most recent fMRI studies in MS have been based on larger and more strictly selected groups of patients than those used in the seminal studies. These studies have investigated the brain patterns of cortical activations during the performance of a number of motor, visual, and cognitive tasks in patients with all the major clinical phenotypes of the disease. One of the most solid conclusions that can be drawn from these fMRI studies is that cortical reorganization does occur in patients with MS. Indeed, the correlation between various measures of structural MS damage and the extent of the cortical activation suggests an adaptive role for cortical changes in contributing to clinical recovery and maintaining a normal level of functioning in patients with MS, despite the presence of irreversible axonal/neuronal loss.

Visual System

The method usually applied to investigate the visual system consists of the application of an 8-Hz photic stimulation to one or both eyes [14, 18, 19-21]. A study of the visual system [18] in patients who had recovered from a single episode of acute unilateral optic neuritis demonstrated that these patients, relative to healthy volunteers, had extensive activation of the visual network, including the claustrum, lateral temporal and posterior parietal cortices, and thalamus, in addition to the primary visual cortex, when the clinically affected eye was studied (Fig. 1). Conversely, when the unaffected eye was stimulated, only activations of the visual cortex and the right insula/claustrum were observed (Fig. 1). Furthermore, a

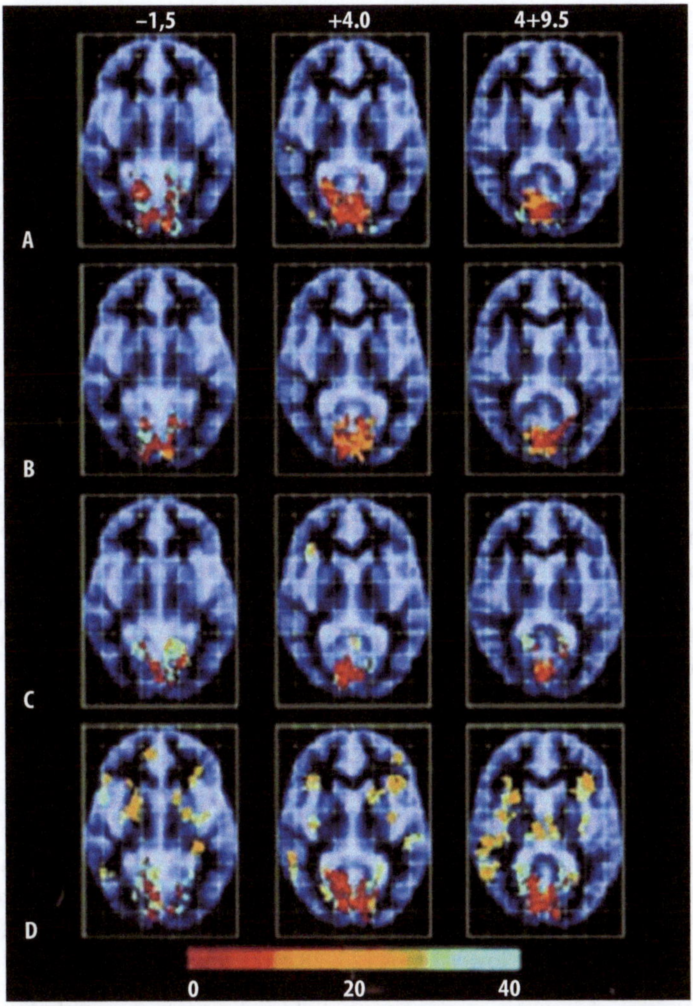

Fig. 1. Brain activation maps from seven control subjects (**A**, **B**) and seven patients with MS (**C**, **D**) showing areas of significant response to monocular photic stimulation compared with binocular darkness. In the control subject group, there is activation only in the visual cortex, bilaterally, after stimulation of either the left (**A**) or the right (**B**) eye. In the patient group, after stimulation of the unaffected eye there is an additional focus of activation in right insula-claustrum (**C**). In the patient group, after stimulation of the affected eye there is additional activation of a network of multimodal processing areas including bilateral insula-claustrum, lateral temporal cortex, posterior parietal cortex, thalamus, and corpus striatum (**D**). Note that the activations are color-coded according to their t values. Images are in radiological convention. Riproduced with permission from [18]

strong correlation was found in these patients between the volume of the extraoccipital activation and the latency of the visual evoked potential (VEP) P100, suggesting that the functional reorganization of the cortex might represent an adaptive response to a persistently abnormal visual input. Rombouts et al. [14] also examined the visual system using fMRI and showed that patients with established MS with unilateral optic neuritis activated a smaller volume of the visual cortex after stimulation of either the affected or the unaffected eye, compared with healthy individuals. On average, patients with optimal clinical recovery showed increased visual cortex activation in comparison with those with poor or no recovery, although activation remained reduced compared with controls [14]. A subsequent study of nine patients with previous optic neuritis confirmed these findings [19] and showed that patients with optic neuritis not only have a low activation of the primary visual cortex but also have a reduced fMRI signal change in this region, suggestive of an abnormality of the synaptic input.

To clarify how early cortical reorganization occurs in the course of MS, and to determine its role in limiting the clinical outcome of the disease, other studies used fMRI to assess the visual system of patients with a single episode of optic neuritis [20, 21]. Toosy et al. [20] replicated the study of Werring et al. [18] using a longer photic stimulation epoch to clarify the nature of the abnormal extraoccipital response observed. The results of this study confirmed the original finding, thus suggesting that cortical functional changes might have a role in compensating for a persistently disordered visual input [20]. Russ et al. [21] used fMRI and VEP to monitor the functional recovery after an acute unilateral optic neuritis. They found a strong relationship between fMRI and P100 latency, suggesting that fMRI might contribute to the assessment of the temporal evolution of the visual deficits during MS recovery or therapy [21].

Motor System

The investigation of the motor system in patients with MS has mainly focused on the analysis of the performance of simple motor tasks with the dominant right upper limbs [15-17, 22-36]. Such tasks were either self-paced or paced by a metronome. A few studies have assessed the performance of simple motor tasks with the dominant right lower limbs [25, 28, 34], while even fewer studies have investigated the performance of more complex tasks, including phasic movements of dominant hand and foot [28, 34] or object manipulation [37].

An altered brain pattern of movement-associated cortical activations, characterized by an increased recruitment of the contralateral primary sensorimotor cortex (SMC) during the performance of simple tasks [24, 28] and by the recruitment of additional "classical" and "higher-order" sensorimotor areas during the performance of more complex tasks [28] has been demonstrated in patients with clinically isolated syndrome (CIS) suggestive of MS. In these patients, the extent of functional cortical changes has been related to the severity of brain axonal damage/dysfunction, measured using whole-brain proton MR spectroscopy (^{1}H-MRS) [24]. This finding suggests that functional cortical reorganization might contribute to the maintenance of normal functional capacities from the earliest clinical stage of MS [24]. The clinical and conventional MRI follow-up of the patients of these two studies has shown that, at disease onset, CIS patients with a subsequent evolution to clinically definite MS tend to recruit a more widespread sensorimotor network than those without short-term disease evolution [29]. These findings suggest that, in CIS patients, the extent of early cortical reorganization following tissue injury might be a factor associated with a different disease evolution. This would support the notion that whereas increased recruitment of widespread sensorimotor network contributes to limiting the impact of structural damage during the course of MS, its early activation could be counterproductive, as it might result in an early exhaustion of the adaptive properties of the brain. This notion is also supported by studies on stroke patients, where a persistent overactivation and overrecruitment of a widespread cortical network has been related to an unfavorable clinical outcome [38].

Pantano et al. [30] studied a group of patients with early MS [39] and a previous episode of hemiparesis. Using a simple motor task, they found that patients had increased activations of several cortical areas mainly located in the ipsilateral hemisphere. Since the patients included in this study had had a previous episode of hemiparesis, interference of task performance on the results [30] could not be ruled out completely. As a consequence, these authors extended their investigations to a group of patients who had experienced a previous episode of optic neuritis [31] as their first clinical manifestation. They found that, during right-hand movement, cortical activations in patients with optic neuritis were mainly located in the contralateral hemisphere, whereas in patients with a previous episode of hemiparesis, increased motor system recruitment was bilateral, with a marked activation of areas located in the ipsilateral hemisphere. The extent of the ipsilateral activations in these patients was correlated with the presence of lesions in the corticospinal tracts [31].

While performing a simple motor task, prominent changes were found in the recruitment of the contralateral and ipsilateral primary SMC and supplementary motor area (SMA) during recovery from an acute relapse in a patient with relapsing/remitting (RR) MS [16]. A strong correlation was also found between the extent of functional cortical changes and N-acetylaspartate (NAA) levels, indicating that dynamic reorganization of the motor cortex can occur in response to an axonal injury associated with MS activity [16].

In patients with clinically stable MS with a varying degree of hand motor impairment, Lee et al. [15] found that the pattern of cortical activations during a simple motor task was significantly different from that of healthy controls. In patients with MS, the increase in the ipsilateral motor cortex activation was significantly related to the T2 lesion load in the contralateral hemisphere. Similarly, by using ^{1}H-MRS, Reddy et al. [17] found that increased activation of the ipsilateral SMC during finger movement was strongly associated with decreased NAA concentration in the brain. These findings suggest that with brain tissue damage, and in particular injury of the motor pathways, there is an increased recruitment of a widespread sensorimotor network, which is likely to contribute to reducing the clinical manifestations of the disease. This has indeed been confirmed in recent work by Rocca et al. [32], who assessed the influence of tissue damage of the brain portion of the left pyramidal tract on the corresponding movement-associated patterns of cortical activation in a sample of 76 right-handed patients with MS when performing a simple motor task with their fully normal-functioning right upper limbs. Lesions along the left pyramidal tract were identified in 43 patients who had more significant activations of the contralateral primary SMC and secondary sensorimotor cortex (SII), inferior central sulcus, and cingulated motor area (CMA), when compared to patients without corticospinal tract lesions. They also showed more significant activations of several regions of the ipsilateral hemisphere, including the primary SMC and the precuneus (Fig. 2). While the T2 lesion load of the corticospinal tract in these patients was correlated with the extent of activation of the contralateral primary SMC, no correlations were found between the extent of activations and the severity of intrinsic lesion damage, as well as with left corticospinal tract NAWM damage.

Despite the presence of irreversible axonal damage [40, 41], quite a large number of patients with RRMS experience relapses and accumulation of an MRI lesion burden without being left with any major residual neurological deficits. To determine whether adaptive cortical reorganization has a role in limiting the impact of irreversible tissue loss in these patients, a study investigated the brain pattern of cortical activations during the performance of a simple motor task in non-disabled patients with RRMS [23]. Compared with healthy controls, these patients showed increased activations of several regions mainly located in the contralateral cerebral hemisphere, including the primary SMC, the SII, and the intraparietal sulcus. In addition, they also showed increased bilateral activations of the SMA and the CMA. Strong correlations were found between the extent of fMRI activations and several magnetization transfer (MT) and diffusion tensor (DT) MRI metrics of structural brain damage. This study demonstrated that cortical activation occurs over a rather diffuse sensorimotor network in non-disabled patients with RRMS, and provided additional evidence showing that an increased recruitment of this cortical network contributes to limiting the functional impact of MS damage associated with macroscopic lesions and subtle normal-appearing brain tissue (NABT) changes. Interestingly, in a group of patients with clinical characteristics similar to those recruited in the previous study, but who complained of fatigue, a reduced activation of a complex movement-associated cortical-subcortical network, including the cerebellum, the rolandic operculum, the thalamus, and the middle frontal gyrus, was detected when compared to matched non-fatigued MS patients [42]. A strong correlation between the reduction of thalamic activity and the clinical severity of fatigue was also found [42], which indicates that a "pseudo-reduction" of brain functional recruitment might be associated with the appearance of MS symptomatology. The notion that fatigue in MS is related to the dysfunction of specific brain networks has been

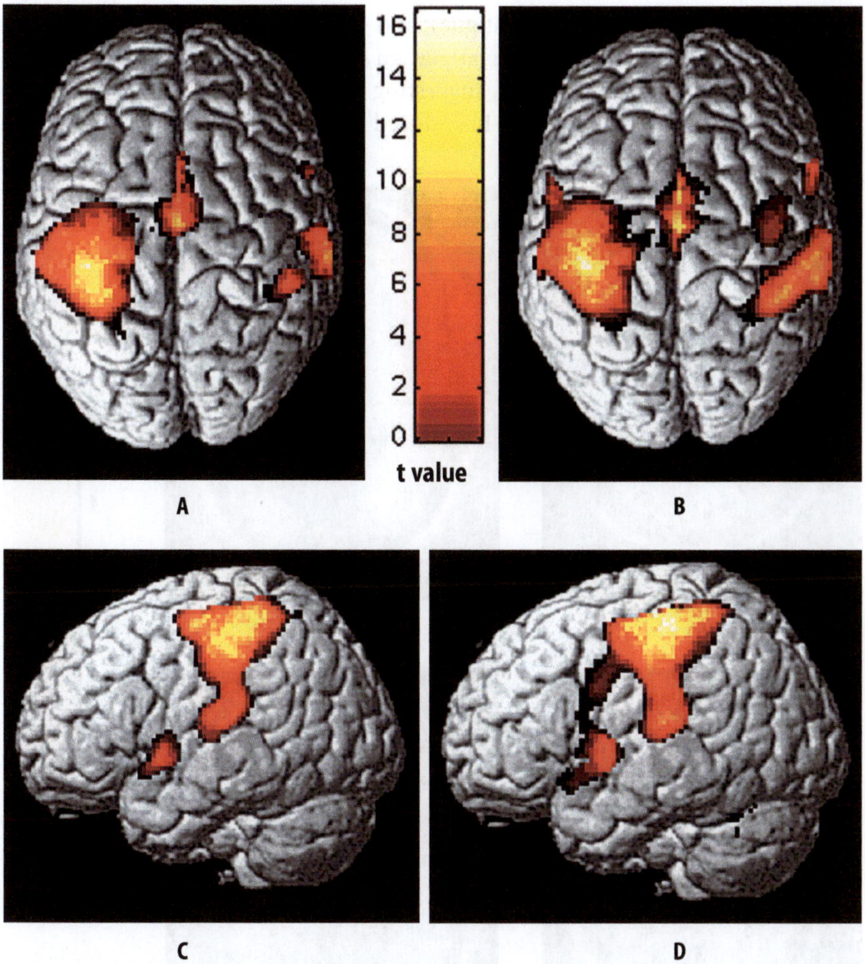

Fig. 2. Brain patterns of cortical activations on a rendered brain from MS patients without (**A**, **C**) and with (**B**, **D**) lesions in the left corticospinal tract, during the performance of a simple motor task with the clinically unimpaired and fully normal functioning, dominant right hand. In patients with corticospinal tract lesions, a more bilateral pattern of activations is visible. Note that the activations are color-coded according to their t values. Images are in neurological convention. (See [32]for further details)

supported by the findings of a recent study, which showed reversible changes in the movement-associated brain pattern of cortical activations in RRMS patients complaining of fatigue after interferon β-1a (IFNβ-1a) administration [43]. In this study, the pattern of cortical activations during the performance of a simple motor task was assessed in a group of 10 patients treated with weekly injection of IFNβ-1a who did not complain of fatigue after treatment administration, and in 12 patients with reversible fatigue after the same treatment was given. fMRI examinations were performed on the same day as IFNβ-1a injection (no fatigue), the day after IFNβ-1a injection (fatigue), and 4 days after IFNβ-1a injection (no fatigue). In patients with reversible fatigue, an abnormal recruitment of the frontothalamic circuitry was observed, suggesting that the dysfunction of this pathway may be associated with IFNβ-1a-induced fatigue in MS patients (Fig. 3).

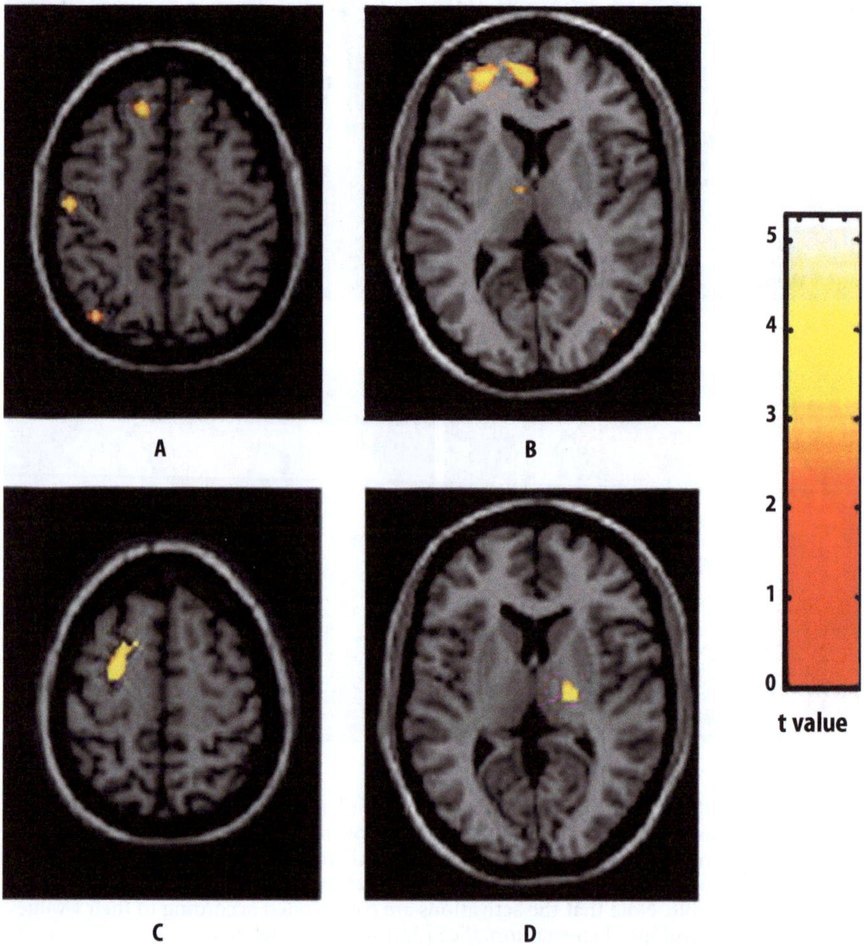

Fig. 3. Relative cortical activations of MS patients with reversible fatigue after IFNβ-1a injection during the performance of a simple motor task with their clinically unimpaired and fully normal functioning, dominant right hands. At entry (**A, B**) (when they did not complain of fatigue), compared with MS patients without reversible fatigue, these patients showed an increased recruitment of the contralateral primary SMC (**A**), the thalamus (**B**), the superior frontal sulcus (**A, B**), and the cingulate motor area (**B**). At time 1 (**C, D**) (when fatigue was present), compared with MS patients without fatigue, these patients showed increased recruitment of the ipsilateral thalamus (**D**), and contralateral middle frontal gyrus (**C**). Note that the color-coded activations have been superimposed on a high-resolution T1-weighted scan obtained from a single, healthy subject and normalized into standard statistical parametric mapping space (neurological convention). From [43], with permission

Since subtle damage of the NABT is an established feature of RRMS [44], the impact of NABT damage, assessed using DT MRI, on movement-associated fMRI activations in patients with RRMS and nonspecific T2-weighted abnormalities on conventional MR images of the brain (i.e., less than three T2 lesions) has been investigated [26]. To assess the role of the non-dominant hemisphere and interhemispheric connections on the dominant hand movement-associated pattern of activations, patients were studied during the performance of simple motor tasks with the dominant and the non-dominant hand. Compared with healthy volunteers, during the performance of a simple motor task with their dominant hand, MS patients showed increased activation of several areas of the sensorimotor network, including the ipsilateral SMA, the ipsilateral superior frontal sulcus, and the contralateral thalamus. The increased recruitment of the SMA was correlated with the peak height and position of the mean diffusivity histogram of the GM. In contrast, when compared with MS patients, healthy subjects showed an increased activation of the ipsilateral primary SMC. When the results obtained from the analysis of the two tasks (i.e., with the dominant and the non-dominant hand) were combined, patients with MS showed an increased activation of the SMA and a reduced activation of the ipsilateral SMC. The most probable explanation for these findings might be an enhanced "transcallosal inhibition," a mechanism that has been postulated to be responsible for the control of homologous hand muscle during unilateral movements. The role of the corpus callosum (CC) in interhemispheric connectivity and in eliciting functional cortical changes has been underpinned by a study by Lowe et al. [33], who showed, by measuring low-frequency BOLD fluctuations, reduced functional connectivity between the right and the left hemisphere primary motor cortices in patients with MS, and by a study of Lenzi et al. [45], who found a relationship between damage of the CC, measured using DT MRI, and the activation of the ipsilateral primary SMC in patients with RRMS.

In an early study by Reddy et al. [27] aimed at determining whether brain injury and disability lead to distinguishable patterns of activations with hand movement, 14 RRMS patients were classified on the basis of lesion load and hand functional impairment. The first group (n = 6) had low diffuse central brain injury, as assessed from relative NAA concentration and normal hand function; the second group (n = 4) had greater diffuse central brain injury and normal hand function; the third group (n = 4) had greater diffuse central brain injury and impaired hand function. The 'injury contrast' (i.e., second vs first group) showed increased activation of the SMA and premotor cortex in patients of the second group, whereas the 'impairment contrast' (i.e., third vs second group) showed increased activation of the SMC and of the parietal cortex, bilaterally, in patients of the third group. Movement-associated cortical changes, characterized by the activation of highly specialized cortical areas, have also been described in patients with secondary progressive (SP) MS during the performance of simple motor tasks [25]. Such cortical changes have been related not only to the extent of T2-visible lesions, but also to the severity of NAWM and GM damage.

The results of all these studies suggest that there may be a "natural history" of the functional reorganization of the cerebral cortex in MS patients, which might be characterized at the beginning of the disease, by an increased recruitment of those areas "normally" devoted to the performance of a given task, such as the primary SMC and the SMA in the case of a motor task. At a later stage, bilateral activation of these regions is first seen, followed by a widespread recruitment of additional areas, which are usually activated in normal people when performing novel/complex tasks. This notion has been supported by the results of a recent study [37], which has provided a direct demonstration that MS patients, during the performance of a simple motor task, activate some regions, which are part of a frontoparietal circuit, whose recruitment occurs typically in healthy subjects during object manipulation. The concept that movement-associated cortical reorganization varies across patients at different stages of the disease has been also shown by a recent fMRI study of patients with different disease phenotypes [46]. The study compared data from 16 patients with a CIS suggestive of MS, 14 with RRMS and no disability, 15 with RRMS and mild clinical disability, and 12 with SPMS. Patients performed a simple motor task with their unimpaired dominant hand during fMRI. CIS patients had an increased activation of the contralateral primary SMC when compared to those with RRMS and no disability, whereas patients with RRMS and no disability had an increased activation of the SMA when compared to those with a CIS (Fig. 4). Patients with RRMS and no disability had an increased activation of the primary SMC, bilaterally, and ipsilateral SMA when compared to patients with RRMS and mild clinical disability. Conversely, patients with RRMS and mild clinical disability had increased activation of the contralateral SII, inferior frontal gyrus, and ipsilateral precuneus. Patients with RRMS and mild clinical disability had increased activation of the contralateral thalamus and ipsilateral SII when compared to those with SPMS. However, patients with SPMS showed increased activation of the inferior frontal gyrus, bilaterally, middle frontal gyrus, bilaterally, contralateral precuneus, and ipsilateral CMA, and inferior parietal lobule. The study suggests that early in the disease course more areas typically devoted to motor tasks are recruited, then a bilateral activation of these regions is seen, and late in the disease course, the areas that healthy people recruit to perform novel or complex tasks are activated [46].

In a longitudinal fMRI study by Pantano et al. [47], a group of MS patients and age-matched controls performed two fMRI scans investigating simple motor function with a time interval of 15-26 months. Patients exhibited greater bilateral activations than controls in both fMRI studies (Fig. 5). While no significant differences between the two fMRI scans were observed in controls, patients showed, at follow-up, a reduction in the functional activity of the ipsilateral SMC and the contralateral cerebellum (Fig. 5). Moreover, activation changes in ipsilateral motor areas correlated inversely with age, extent, and progression of T1 lesion load, and occurrence of a new relapse, suggesting that younger patients with less structural brain damage and benign clinical course demonstrate brain plasticity that follows

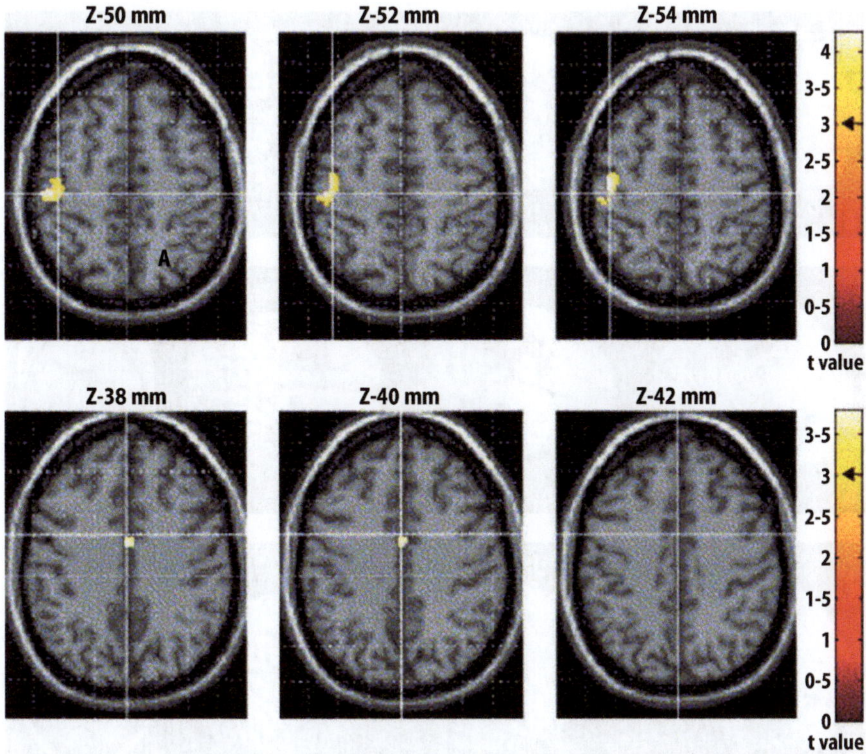

Fig. 4. Comparisons of patients with a CIS suggestive of MS and patients with RRMS and no disability during a simple right-hand motor task. Patients with a CIS had increased activation of the contralateral primary sensorimotor cortex when compared with patients with RRMS and no disability (*top row*). Patients with RRMS and no disability had more significant activation of the supplementary motor area, bilaterally, when compared with patients with a CIS (*bottom row*). Images are color-coded for activation and *arrows* show *t* cut-off values. Activations were superimposed on a high-resolution T1-weighted scan obtained from one healthy individual and normalized into a standard statistical parametric mapping space (neurological convention). Reproduced with permission from [46]

a more lateralized pattern of brain activation. The tendency towards a normalization of brain functional activity seems to be hampered in older patients and in those developing relapses or new irreversible brain damage [47].

At present, few studies have investigated the role of cortical plasticity in limiting the functional consequences of tissue damage in patients with primary progressive (PP) MS [22, 34, 48]. In one study, the patterns of brain activations following simple and complex motor tasks were assessed in PPMS patients with variable degrees of motor impairment, and in controls matched for sex and age [34]. Significant cortical activation changes during the performance of both the simple and the complex motor tasks were found in these patients, who showed different patterns of cortical activations with the varying degrees of their clinical impairment. During the performance of a simple motor task with the clinically

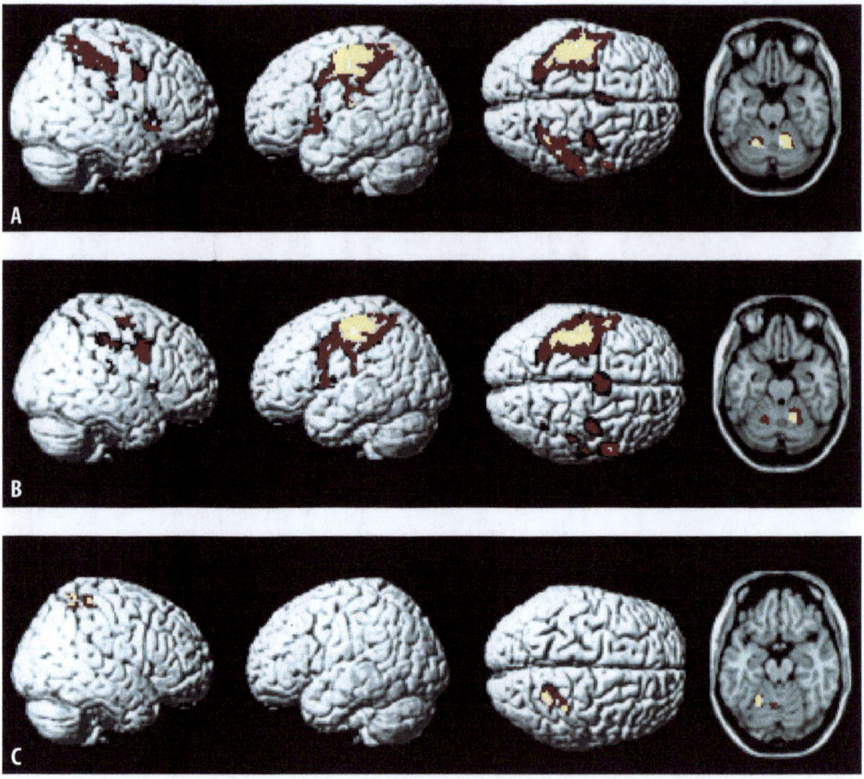

Fig. 5. Group maps generated from random effect analysis showing task-related activation at baseline (A), task-related activation at follow-up (B), and task-related activity decrease between the two fMRI studies during right-hand movement (C) in 18 patients with MS. Areas of decreased activity (baseline > follow-up) (C) included the right (ipsilateral) sensorimotor cortex and the left (contralateral) cerebellum. Note that the activations are color-coded according to their t values. Images are in neurological convention. Reproduced with permission from [47]

unaffected limbs, patients with PPMS had more significant activations of the contralateral SMA, the SII, bilaterally, and several regions of the frontal and temporal lobes (e.g., insula and superior temporal gyrus) in comparison with healthy volunteers. These findings indicate a marked cortical reorganization taking place in brain regions involved in different phases of movement planning and execution. Interestingly, the observed pattern of cortical activations involved a widespread network usually considered to function in motor, sensory, and multimodal integration processing (i.e., the frontal and temporal cortex and the insula) [49]. During the performance of a simple task with an affected limb, patients with PPMS showed increased activations of the ipsilateral CMA and the postcentral gyrus. Since CMA activation in healthy subjects has been related to the learning of new motor tasks, and since it is also thought to reflect task difficulty [50], a possible explanation for the increased CMA recruitment found in patients with PPMS might be that the patients tended to perceive the simple experimental task as a sort of a novel task, which, as a consequence, had to be "relearned." Finally, during the execution of a more complex task, patients with PPMS showed increased activations of the ipsilateral thalamus, the middle frontal gyrus, bilaterally, and several other sensory regions, including an area located in the visual cortex. Visual-sensory interactions are known to occur in humans [51] and, again, they might be enhanced in patients with PPMS in an attempt to compensate for the functional impairment secondary to subcortical WM damage. The notion that multimodal integration areas might have a critical role in PPMS patients has been strengthened by another study, which showed increased activations of these regions in PPMS patients, in comparison with healthy controls, during passive movements [48]. In another fMRI study of a motor task, the hypothesis tested was that if cortical adaptive responses have the potential to limit the accumulation of disability in patients with PPMS after tissue injury, the extent of such changes should be greater with increasing MS-related pathology [22]. Consistent with this hypothesis, strong correlations between the extent of the fMRI activations of several sensorimotor areas and several MT or DT MRI metrics of structural damage of the brain and the cervical cord were found. This suggests that not only brain damage, but also cervical cord involvement, can cause adaptive cortical reorganization with the potential to limit the functional consequences of MS-related structural damage.

To assess the role of cervical cord pathology in isolation in eliciting brain functional reorganization, movement-associated cortical activations have been investigated in patients with a previous episode of isolated acute myelitis of probable demyelinating origin and lack of brain damage [35]. In this study, the relationship between the extent of cortical reorganization and the extent of cord pathology, measured using MT MRI, was also evaluated. Compared with healthy volunteers, patients with myelitis had increased recruitment of the ipsilateral primary SMC, SMA, and middle frontal gyrus. Several strong correlations were found between the extent of fMRI activations and MT MRI metrics of cervical cord damage. Recently, it has also been shown that patients with isolated myelitis have different patterns of movement-associated cortical activations according to the level of cord damage [52]. The relation between the presence and extent of spinal cord injury and an altered brain pattern of movement-associated cortical activations has also been shown in patients with Devic's neuromyelitis optica [36].

A recent study used fMRI to analyze how the motor network responds to motor training in MS patients with mild motor impairment of the right upper extremity [53]. Before training, MS patients had a more prominent activation of the contralateral dorsal premotor cortex (BA6) during thumb movements, when compared with controls. After training, unlike the control group, MS patients did not exhibit task-specific reductions in activation of the contralateral primary somatosensory, motor and adjacent parietal association (BA40) cortices (Fig. 6). The absence of training-dependent reductions in activations supports the notion

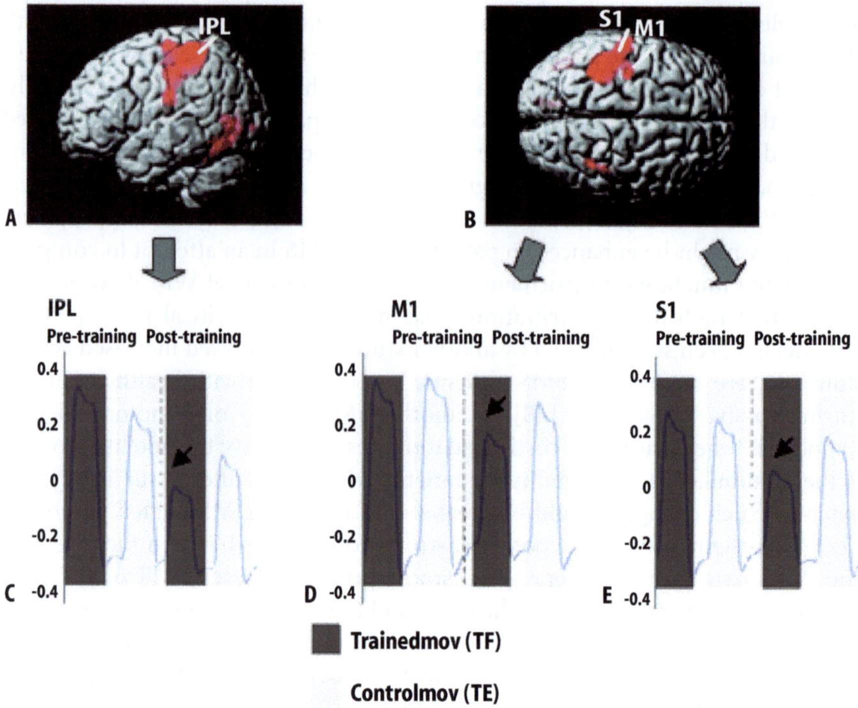

Fig. 6. Three-dimensional rendering (**A**, **B**) of task-specific reductions of activations in contralateral inferior parietal lobule (IPL) (BA 40; **A** and **C**), primary motor cortex (M1) (**B** and **D**), and primary somatosensory cortex (S1) (**B** and **E**) in healthy volunteers after training. **C**, **D**, and **E** illustrate average signal intensities in voxels representing cluster maxima in IPL (BA 40), M1, and S1 for 30-s blocks of thumb flexion (trained movement, *gray bars*), extension (control movement, *blue bars*), and rest. Note that training led to a decrease in signal intensity predominantly for the trained movement (*arrows*). Images are in neurological convention. Reproduced with permission from [53]

that a decreased capacity to optimize recruitment of the motor network with practice is present in MS patients. Further studies are now warranted in order to better understand training-dependent cortical plasticity in MS, which might help to optimize therapies and prolong preservation of motor functions.

Cognition

Recent fMRI studies have suggested that functional cortical changes might have an adaptive role in limiting MS-related cognitive impairment [54-66]. Therefore, brain "plasticity" might, in part, explain the weak relationship found in MS between neuropsychological deficits and conventional measures of disease burden [67].

Several cognitive domains have been investigated in MS patients, using fMRI. Among these cognitive domains, working memory has been the most extensively studied by means of the Paced Auditory Serial Addition Test (PASAT) and the Paced Visual Serial Addition Task (PVSAT) [54, 55-57, 61, 65], the n-back task [60, 62-64, 66], or a task adapted from the Sternberg paradigm [58]. Additional cognitive domains including attention [59] and planning [68] have also been investigated. The PASAT and its visual version, the PVSAT, involve working memory as well as sustained attention, information processing speed, and simple calculation. Using the PASAT, Audoin et al. [57] showed that relative to controls, patients with CIS suggestive of MS had significantly greater activations of the right frontopolar cortex, the bilateral lateral prefrontal cortices, and the right cerebellum. Both groups had similar performances on the task, confirming that compensatory cortical activations at the earliest stage of MS reflect active processes of neuroplasticity that are likely to contribute to reducing the cognitive outcome of brain MS pathology. During performance on the PVSAT, Staffen et al. [54] observed differences in the patterns of brain activations between RRMS patients and healthy controls, despite the fact that patients had intact performance on the task. These findings were again interpreted as compensatory mechanisms through neuronal plasticity during the early stages of the disease.

Other authors have investigated the relationship between functional adaptive changes and the extent of overall tissue damage on quantitative [69] and conventional MRI [61]. Audoin et al. [69] compared 18 patients with a CIS suggestive of MS with 18 controls on the PASAT. Structural abnormalities were revealed in the NAWM and GM of patients by means of MT ratio (MTR) histogram analyses. As a whole, patients performed more poorly than controls, and their PASAT scores correlated with the mean NAWM MTR, but not with the GM MTR. Relative to control subjects, patients showed increased activations of the BA45, bilaterally, and the right BA44. In patients, the activation of the right BA45 was inversely correlated with the mean NAWM MTR and the peak position of the GM MTR histograms. In a study by Mainero et al. [61], patients with RRMS and no, or only mild, deficits at neuropsychological testing were compared to healthy subjects during the performance of attention and memory tasks. During both the PASAT, and a recall task, patients exhibited significantly greater brain activations than controls and recruited additional brain areas. In patients, task-related activations during performance of both tasks correlated with T2 lesion load, and were greater in patients with better cognitive functioning than in those with impaired cognition.

Using PASAT, altered functional [55] and effective [56] connectivities of the working memory network have been demonstrated in patients at the very early clinical stage of MS. Altered functional connectivity was found to be related to the extent of diffuse WM changes, assessed using MTR, and T2 lesion load [55].

The n-bask task is another paradigm that has been commonly used to study working memory in MS. Wishart et al. [64] showed that, relative to healthy controls, patients with mild RRMS showed shifts in brain activation patterns within (i.e., prefrontal and parietal regions) and outside (i.e., bilateral medial frontal, cingulate, parietal, bilateral middle temporal, and occipital regions) the typical components of the working memory circuitry. Using a 2-back verbal working memory task, Sweet et al. [62] showed that a shift toward increased activation of regions related to sensorimotor functions and anterior attentional/executive components of the verbal working memory system contributed to sustain normal performance of a challenging verbal working memory task among high-functioning MS patients. Posterior memory storage systems appeared unaffected, while portions of the visual processing and subvocal rehearsal systems were found to be less active. Although a shift in neural activity was noted relative to healthy controls, deviation from regions normally involved in verbal working memory function was not observed in this patient sample. Moreover, compensatory activity occurred predominantly in regions associated with verbal working memory, and this may decline relative to controls as task demands increase, which might explain why MS patients' performance decreased as a function of effort on neuropsychological tests [63].

In a recent fMRI study conducted at 3.0 T [66], working memory was investigated with an n-back task and functional connectivity analysis in a group of 21 RRMS patients and 16 age- and sex-matched healthy control subjects. With similar performances on the task, activations were found in similar regions for both groups. However, patients had relatively reduced activations of the superior frontal and anterior cingulate gyri. Patients also showed a variable, but generally substantially smaller, increase in activation than healthy controls with greater task complexity, depending on the specific brain regions assessed. These findings suggest that, despite the recruitment of similar brain regions for the task in both groups, patients have a reduced functional reserve for cognition relevant to memory. The functional connectivity analysis revealed greater correlations between right dorsolateral prefrontal and superior frontal/anterior cingulate activations in controls, and correlations between activations in the right and left prefrontal cortices in patients, suggesting that altered interhemispheric interactions between dorsal and lateral prefrontal regions may be an additional adaptive mechanism that could limit the clinical manifestations of the disease distinct from recruitment of novel processing regions [66].

Conclusions

Taken together, fMRI studies demonstrate the potential of this technique in providing important insights into the mechanisms of cortical reorganization following MS-related tissue injury. This should improve our understanding of the factors associated with the accumulation of irreversible disability in MS. Although the role of cortical reorganization in limiting the functional impact of MS-related structural damage has still not been proved conclusively, the available data support the hypothesis that cortical adaptive responses may have an important role in compensating for MS-related irreversible tissue damage, such as axonal loss. Thereby, it can be concluded that the presently available fMRI data suggest that the rate of accumulation of disability in MS might be a function not only of tissue loss but also of the progressive failure of the adaptive capacity of the brain.

References

1. Filippi M, Rocca MA, Comi G (2003) The use of quantitative magnetic-resonance-based techniques to monitor the evolution of multiple sclerosis. Lancet Neurol 2:337-346
2. Inglese M, Benedetti B, Filippi M (2005) The relation between MRI measures of inflammation and neurodegeneration in multiple sclerosis. J Neurol Sci 15:15-19
3. Bjartmar C, Trapp BD (2001) Axonal and neuronal degeneration in multiple sclerosis: mechanisms and functional consequences. Curr Opin Neurol 14:271-278
4. Bjartmar C, Wujek JR, Trapp BD (2003) Axonal loss in the pathology of MS: consequences for understanding the progressive phase of the disease. J Neurol Sci 15:165-171
5. Waxman SG, Ritchie JM (1993) Molecular dissection of the myelinated axon. Ann Neurol 33:121-136
6. Filippi M, Rocca MA (2003) Disturbed function and plasticity in multiple sclerosis as gleaned from functional magnetic resonance imaging. Curr Opin Neurol 16:275-282
7. Ogawa S, Menon RS, Tank DW et al (1993) Functional brain mapping by blood oxygenation level-dependent contrast magnetic resonance imaging: a comparison of signal characteristics with a biophysical model. Biophys J 64:803-812
8. Jueptner M, Weiller C (1995) Review: does measurement of regional cerebral blood flow reflect synaptic activity? Implications for PET and fMRI. Neuroimage 2:148-156
9. Malonek D, Grinvald A (1996) Interactions between electrical activity and cortical microcirculation revealed by imaging spectroscopy: implications for functional brain mapping. Science 272:551-554
10. Vanzetta I, Grinvald A (1999) Increased cortical oxidative metabolism due to sensory stimulation: implications for functional brain imaging. Science 286:1555-1558
11. Bandettini PA, Jesmanowicz A, Wong EC, Hyde JS (1993) Processing strategies for time-course data sets in functional MRI of the human brain. Magn Reson Med 30:161-173
12. Worsley KJ, Friston KJ (1995) Analysis of fMRI time-series revisited - again. Neuroimage 2:173-181
13. Clanet M, Berry I, Boulanouar K (1997) Functional imaging in multiple sclerosis. Int MS J 4:26-32
14. Rombouts SA, Lazeron RH, Scheltens P et al (1998) Visual activation patterns in patients with optic neuritis: an fMRI pilot study. Neurology 50:1896-1899
15. Lee MA, Reddy H, Johansen-Berg H et al (2000) The motor cortex shows adaptive functional changes to brain injury from multiple sclerosis. Ann Neurol 47:606-613
16. Reddy H, Narayanan S, Matthews PM et al (2000) Relating axonal injury to functional recovery in MS. Neurology 54:236-239
17. Reddy H, Narayanan S, Arnoutelis R et al (2000) Evidence for adaptive functional changes in the cerebral cortex with axonal injury from multiple sclerosis. Brain 123:2314-2320
18. Werring DJ, Bullmore ET, Toosy AT et al (2000) Recovery from optic neuritis is associated with a change in the distribution of cerebral response to visual stimulation: a functional magnetic resonance imaging study. J Neurol Neurosurg Psychiat 68:441-449

19. Langkilde AR, Frederiksen JL, Rostrup E, Larsson HB (2002) Functional MRI of the visual cortex and visual testing in patients with previous optic neuritis. Eur J Neurol 9:277-286
20. Toosy AT, Werring DJ, Bullmore ET et al (2002) Functional magnetic resonance imaging of the cortical response to photic stimulation in humans following optic neuritis recovery. Neurosci Lett 330:255-259
21. Russ MO, Cleff U, Lanfermann H et al (2002) Functional magnetic resonance imaging in acute unilateral optic neuritis. J Neuroimaging 12:339-350
22. Filippi M, Rocca MA, Falini A et al (2002) Correlations between structural CNS damage and functional MRI changes in primary progressive MS. Neuroimage 15:537-546
23. Rocca MA, Falini A, Colombo B et al (2002) Adaptive functional changes in the cerebral cortex of patients with non-disabling MS correlate with the extent of brain structural damage. Ann Neurol 51:330-339
24. Rocca MA, Mezzapesa DM, Falini A et al (2003) Evidence for axonal pathology and adaptive cortical reorganization in patients at presentation with clinically isolated syndromes suggestive of multiple sclerosis. Neuroimage 18:847-855
25. Rocca MA, Gavazzi C, Mezzapesa DM et al (2003) A functional magnetic resonance imaging study of patients with secondary progressive multiple sclerosis. Neuroimage 19:1770-1777
26. Rocca MA, Pagani E, Ghezzi A et al (2003) Functional cortical changes in patients with MS and non-specific conventional MRI scans of the brain. Neuroimage 19:826-836
27. Reddy H, Narayanan S, Woolrich M et al (2002) Functional brain reorganization for hand movement in patients with multiple sclerosis: defining distinct effects of injury and disability. Brain 125:2646-2657
28. Filippi M, Rocca MA, Mezzapesa DM et al (2004) Simple and complex movement-associated functional MRI changes in patients at presentation with clinically isolated syndromes suggestive of MS. Hum Brain Mapp 21:108-117
29. Rocca MA, Mezzapesa DM, Ghezzi A et al (2005) A widespread pattern of cortical activations in patients at presentation with clinically isolated symptoms is associated with evolution to definite multiple sclerosis. Am J Neuroradiol 26:1136-1139
30. Pantano P, Iannetti GD, Caramia F et al (2002) Cortical motor reorganization after a single clinical attack of multiple sclerosis. Brain 125:1607-1615
31. Pantano P, Mainero C, Iannetti GD et al (2002) Contribution of corticospinal tract damage to cortical motor reorganization after a single clinical attack of multiple sclerosis. Neuroimage 17:1837-1843
32. Rocca MA, Gallo A, Colombo B et al (2004) Pyramidal tract lesions and movement-associated cortical recruitment in patients with MS. Neuroimage 2004:23:141-147
33. Lowe MJ, Phillips MD, Lurito JT et al (2002) Multiple sclerosis: low-frequency temporal blood oxygen level-dependent fluctuations indicate reduced functional connectivity initial results. Radiology 224:184-192
34. Rocca MA, Matthews PM, Caputo D et al (2002) Evidence for widespread movement-associated functional MRI changes in patients with PPMS. Neurology 58:866-872
35. Rocca MA, Mezzapesa DM, Ghezzi A et al (2003) Cord damage elicits brain functional reorganization after a single episode of myelitis. Neurology 61:1078-1085
36. Rocca MA, Agosta F, Mezzapesa DM et al (2004) A functional MRI study of movement-associated cortical changes in patients with Devic's neuromyelitis optica. Neuroimage 21:1061-1068
37. Filippi M, Rocca MA, Mezzapesa DM et al (2004) A functional MRI study of cortical activations associated with object manipulation in patients with MS. Neuroimage 21:1147-1154
38. Calautti C, Baron JC (2003) Functional neuroimaging studies of motor recovery after stroke in adults: a review. Stroke 34:1553-1566

39. McDonald WI, Compston A, Edan G et al (2001) Recommended diagnostic criteria for multiple sclerosis: guidelines from the International Panel on the diagnosis of multiple sclerosis. Ann Neurol 50:121-127
40. Ferguson B, Matyszak MK, Esiri MM, Perry VH (1997) Axonal damage in acute multiple sclerosis lesions. Brain 120:393-399
41. Trapp BD, Ransohoff R, Rudick R (1999) Axonal pathology in multiple sclerosis: relationship to neurologic disability. Curr Opin Neurol 12:295-302
42. Filippi M, Rocca MA, Colombo B et al (2002) Functional magnetic resonance imaging correlates of fatigue in multiple sclerosis. Neuroimage 15:559-567
43. Rocca MA, Agosta F, Colombo B et al (2007) fMRI changes in relapsing-remitting multiple sclerosis patients complaining of fatigue after IFNβ-1a injection. Hum Brain Mapp [epub 24 Aug 2006; doi:10.1002/hbm.20279]
44. Filippi M, Rocca MA, Comi G (2003) The use of quantitative magnetic-resonance-based techniques to monitor the evolution of multiple sclerosis. Lancet Neurol 2:337-346
45. Lenzi D, Conte A, Mainero C et al (2007) Effect of corpus callosum damage on ipsilateral motor activation in patients with multiple sclerosis: a functional and anatomical study. Hum Brain Mapp [epub 1 Nov 2006; doi 10.1002/hbm.20305]
46. Rocca MA, Colombo B, Falini A et al (2005) Cortical adaptation in patients with MS: a cross-sectional functional MRI study of disease phenotypes. Lancet Neurol 4:618-626
47. Pantano P, Mainero C, Lenzi D et al (2005) A longitudinal fMRI study on motor activity in patients with multiple sclerosis. Brain 128:2146-2153
48. Ciccarelli O, Toosy AT, Marsden JF et al (2006) Functional response to active and passive ankle movements with clinical correlations in patients with primary progressive multiple sclerosis. J Neurol 253:882-891
49. Mesulam MM (1998) From sensation to cognition. Brain 121:1013-1052
50. Rao SM, Binder JR, Bandettini PA et al (1993) Functional magnetic resonance imaging of complex human movements. Neurology 43:2311-2318
51. de Gelder B (2000) Neuroscience: more to seeing than meets the eye. Science 289:1148-1149
52. Rocca MA, Agosta F, Martinelli V et al (2006) The level of spinal cord involvement influences the pattern of movement-associated cortical recruitment in patients with isolated myelitis. Neuroimage 15:879-884
53. Morgen K, Kadom N, Sawaki L et al (2004) Training-dependent plasticity in patients with multiple sclerosis. Brain 127, 2506-2517
54. Staffen W, Mair A, Zauner H et al (2002) Cognitive function and fMRI in patients with multiple sclerosis: evidence for compensatory cortical activation during an attention task. Brain 125:1275-1282
55. Au Duong MV, Audoin B, Boulanouar K et al (2005) Altered functional connectivity related to white matter changes inside the working memory network at the very early stage of MS. J Cereb Blood Flow Metab 25:1245-1253
56. Au Duong MV, Boulanouar K, Audoin B et al (2005) Modulation of effective connectivity inside the working memory network in patients at the earliest stage of multiple sclerosis. Neuroimage 24:533-538
57. Audoin B, Ibarrola D, Ranjeva JP et al (2003) Compensatory cortical activation observed by fMRI during a cognitive task at the earliest stage of MS. Hum Brain Mapp 20:51-58
58. Hillary FG, Chiaravalloti ND, Ricker JH et al (2003) An investigation of working memory rehearsal in multiple sclerosis using fMRI. J Clin Exp Neuropsychol 25:965-978
59. Parry AM, Scott RB, Palace J et al (2003) Potentially adaptive functional changes in cognitive processing for patients with multiple sclerosis and their acute modulation by rivastigmine. Brain 126:2750-2760

60. Penner IK, Rausch M, Kappos L et al (2003) Analysis of impairment related functional architecture in MS patients during performance of different attention tasks. J Neurol 250:461-472
61. Mainero C, Caramia F, Pozzilli C et al (2004) fMRI evidence of brain reorganization during attention and memory tasks in multiple sclerosis. Neuroimage 21:858-867
62. Sweet LH, Rao SM, Primeau M et al (2004) Functional magnetic resonance imaging of working memory among multiple sclerosis patients. J Neuroimaging 14:150-157
63. Sweet LH, Rao SM, Primeau M et al (2006) Functional magnetic resonance imaging response to increased verbal working memory demands among patients with multiple sclerosis. Hum Brain Mapp 27:28-36
64. Wishart HA, Saykin AJ, McDonald BC et al (2004) Brain activation patterns associated with working memory in relapsing-remitting MS. Neurology 62:234-238
65. Chiaravalloti N, Hillary FG, Ricker JH et al (2005) Cerebral activation patterns during working memory performance in multiple sclerosis using fMRI. J Clin Exp Neuropsychol 27:33-54
66. Cader S, Cifelli A, Abu-Omar Y et al (2006) Reduced brain functional reserve and altered functional connectivity in patients with multiple sclerosis. Brain 129:527-537
67. Comi G, Rovaris M, Leocani L et al (2001) Clinical and MRI assessment of brain damage in MS. Neurol Sci 22:S123-S127
68. Lazeron RH, Rombouts SA, Scheltens P et al (2004) An fMRI study of planning-related brain activity in patients with moderately advanced multiple sclerosis. Mult Scler 10:549-555
69. Audoin B, Au Duong MV, Ranjeva JP et al (2005) Magnetic resonance study of the influence of tissue damage and cortical reorganization on PASAT performance at the earliest stage of multiple sclerosis. Hum Brain Mapp 24:216-228

EVALUATION OF MRI OUTCOMES

Chapter 10

Validation of MRI Surrogates

M.P. Sormani, M. Filippi

Introduction

Traditionally, the clinical endpoints used to monitor clinical trials in multiple sclerosis (MS) are the occurrence of relapses (usually expressed as a rate over time), and the degree of disability, most commonly evaluated using the Expanded Disability Status Scale (EDSS) [1].

Magnetic resonance imaging (MRI) is commonly used for studying MS, and MRI-derived endpoints are, at present, largely used as primary measures of outcome in phase-II trials and as secondary measures of outcome in phase-III trials of MS. So far, statisticians working in this field have been asked to assess the precision, accuracy, sensitivity to change, cost-effectiveness, and correlation between clinical and MRI-derived endpoints; however, the central issue that needs to be clarified in order to use MRI in studies of MS as a surrogate outcome for clinical measures still remains controversial [2].

Surrogate endpoints (SE) are defined as instrumental or laboratory variables with the potential to be used in clinical trials instead of the "true" clinical outcome of interest. An SE is useful when it can be measured earlier, more conveniently, or more sensitively than the outcome of interest. Under these circumstances, the use of an SE can result in significant reductions in the duration of the trial and the size of the sample studied. It might also be helpful when trying to anticipate the effect of a given treatment on the management of individual patients.

The criteria that an outcome measure must meet to be considered as a valid SE are stringent and often not fully acknowledged. In this context, the most common misconception is to consider as valid SEs those outcome measures found to be correlated with the "true" clinical outcome. Clearly, the demonstration of such a correlation is a prerequisite for validating surrogacy, but it is not sufficient on its own. The adoption of a valid SE also requires that the effect of a given treatment on the SE allows the prediction of the effect on the outcome of interest. Prentice [3], in a landmark paper, proposed a formal definition of SEs and outlined how potential SEs could be validated using data from individual clinical trials and four operational criteria. To consider a paraclinical measure as a valid SE for a clinical endpoint of interest, these criteria require that a given treatment is effective on the SE (*first criterion*) and on the clinical endpoint of interest (*second criterion*), that the SE and the clinical endpoints are significantly correlated (*third criterion*) and, finally, that the effect of a given treatment on the clinical endpoint is mediated through an effect on the SE (*fourth criterion*) in such a way that the treatment effect "disappears" when adjusting for the SE.

At present, only two validation studies have been performed in MS [4, 5], which have investigated the role of MRI as an SE for clinical variables in relapsing/remitting multiple sclerosis (RRMS) patients treated with glatiramer acetate (GA) and in secondary progressive multiple sclerosis (SPMS) patients treated with interferon β-1b (IFNβ-1b). These studies used data from two randomized trials [6, 7] and the statistical methodology proposed by Prentice [3].

A Brief Introduction to the Prentice Criteria

The Prentice criteria can be formalized by building up appropriate models for the treatment effect, according to the nature of the dependent variable. In general, indicating the treatment variable with Z, the intercept with μ, and the error term with ε for each model, they can be summarized as follows:

1. ***Z* must be prognostic for *S*.** This condition can be tested by assessing the significance of the coefficient α in the following regression model:

$$S_j \mid Z_j = \mu_S + \alpha Z_j + \varepsilon_{Sj}$$

   where α = treatment effect on *S*.

2. ***Z* must be prognostic for *T*.** This condition can be tested by assessing the significance of the coefficient β in the following regression model:

$$T_j \mid Z_j = \mu_T + \beta Z_j + \varepsilon_{Tj}$$

   where β = treatment effect on *T*.

3. ***S* must be prognostic for *T*.** This condition can be tested by assessing the significance of the coefficient γ in the following regression model:

$$T_j \mid S_j = \mu + \gamma S_j + \varepsilon_j$$

   where γ = effect of *S* on *T*.

4. **The full effect of *Z* on *T* must be explained by *S*.** This condition is satisfied when the coefficient β_S in the following regression model becomes equal to zero:

$$T_j \mid Z_j, S_j = \mu_{TS} + \beta_S Z_j + \gamma_Z S_j + \varepsilon_{TSj}$$

   where β_S = treatment effect on *T* adjusted for *S*, and γ_Z = effect of *S* on *T* adjusted for treatment.

Validating Conventional MRI Metrics as Surrogates for Relapses in GA-treated RRMS Patients

The study design of this randomized trial is reported elsewhere [6].

The total number of relapses over the 9-month follow-up period was considered the "true" (T) outcome for this validation study, whereas the number of new enhancing lesions seen on monthly post-contrast T1-weighted scans, the percentage lesion volume change between the baseline and the exit T2-weighted scans and a composite measure of these two quantities were considered as the surrogate (S_1, S_2, S_3) outcomes. The results of this validation study showed that the four Prentice criteria [3] were satisfied. GA significantly reduced the number of relapses: the relative risk (RR) was equal to 0.72 (RR = 0.72; 95% CI = 0.52-0.99), indicating that GA-treated patients experienced about 28% fewer relapses than patients on placebo. According to the second Prentice criterion, this indicates that the validity of the explored MRI metrics for being considered as SEs for clinical relapses could be investigated further.

As regards the number of new enhancing lesions, the first Prentice criterion was also satisfied, since the effect of treatment on this metric was statistically significant: RR = 0.60 (95% CI = 0.44-0.81), indicating that the treatment reduced the mean number of such lesions by about 40%. The third Prentice criterion was also satisfied, since a strong relationship between the number of new enhancing lesions and the number of relapses was found ($P < 0.0001$). Satisfaction of the fourth Prentice criterion requires that the reduction in the number of relapses due to treatment is mediated through a reduction in the number of new enhancing lesions. In other words, the treatment effect on relapses should "disappear" when adjustment for lesion number is applied. A strong decrease in treatment effect on relapses was observed after such an adjustment. Nevertheless, a residual treatment effect was still present: GA-treated patients still experienced about 17% fewer relapses than placebo patients after the adjustment ($P = 0.28$).

As regards the percentage T2 lesion volume change, similar results were obtained in terms of satisfaction of the Prentice criteria. The effect of treatment on the percentage T2 lesion load change was statistically significant: the percentage T2 lesion volume change induced by treatment was about 12% (first criterion). The third criterion was also satisfied, since a strong relationship between the percentage T2 lesion volume change and number of relapses was found ($P = 0.002$). Finally, adjusting for T2 lesion load change (fourth criterion) resulted in a decreased treatment effect on relapses. Nevertheless, a residual treatment effect was still present: GA-treated patients continued to experience about 15% fewer relapses than placebo patients ($P = 0.36$).

The composite MRI score was also different between GA-treated and placebo patients ($P = 0.002$), and was strongly correlated with the number of relapses ($P < 0.0001$). The composite MRI score accounted for a large proportion of the treatment effect on relapses. After adjusting for the composite MRI score, it was found that GA-treated patients experienced about 9% fewer relapses than placebo patients (RR = 0.91; 95% CI = 0.6-1.28).

Validating Conventional MRI Metrics as Surrogates for Clinical Disability in IFNβ-1b-treated SPMS Patients

The study design of this randomized trial is reported elsewhere [7].

The level of disability measured using the EDSS score at study exit (3 years), adjusted for the baseline EDSS, was the primary clinical outcome of this validation study. Two surrogate MRI variables were considered: the number of new and enlarging lesions ("active" T2 lesions) seen on the T2-weighted MRI scans obtained at year 1 (S_1); and the percentage lesion volume change between T2-weighted MRI scans obtained at baseline and year 1 (S_2). A composite MRI score including both these MRI metrics was also considered (S_3).

In the patient sample which entered the analysis, the first Prentice criterion was satisfied, since the effect of treatment on the number of "active" T2 lesions was significant: the exponential of the coefficient representing the treatment effect was 0.40 (95% CI = 0.31-0.52), indicating that the administration of the "active" treatment was associated with a 60% reduction in the mean number of "active" T2 lesions accumulated over the first year. The effect of treatment was also significant in reducing the percentage lesion volume change over the first year: a difference of 7.3% (95% CI = 5.1%-9.4%) was observed in favor of the treated patients during the first year of the study.

In the present sample, the treatment effect on EDSS was also confirmed (second Prentice criterion). The coefficient β, representing the treatment effect, was –0.28 (SEM = 0.08, P = 0.001), which means that, on average, the mean EDSS change over the 3 years of the study was lower in the treated patients by 0.28 points than in those receiving placebo.

The third Prentice criterion was also satisfied, since a weak, but significant, correlation between the number of "active" T2 lesions accumulated over the first year and the EDSS change over the entire follow-up duration was found (r = 0.16, $P < 0.0001$). The same was true for the correlation between the percentage T2 lesion volume change over the first year and EDSS changes over the entire study period (r = 0.13, P = 0.002).

When the treatment effect was adjusted for the number of active T2 lesions, the impact on EDSS at year 3 (corrected for the baseline EDSS) changed from –0.28 EDSS points to –0.16 EDSS points (SEM = 0.09, P = 0.08). A similar finding was seen when considering the percentage T2 lesion volume change over the first year: the residual treatment effect after adjusting for T2 lesion volume change was –0.20 (SEM = 0.09, P = 0.02) (fourth Prentice criterion). When both these MRI variables were included in the model, the residual IFNβ-1b effect was equal to –0.12 (SEM = 0.10, P = 0.28).

Conclusions

These studies, based on data collected in two large-scale MS treatment trials of GA and IFNβ-1b [6, 7], represent a first attempt to achieve a formal validation of MRI metrics as SEs for clinical relapses and EDSS progression in patients with MS.

Although confirmation of these results are needed, the first study [6] indicates that two of the most commonly accepted MRI metrics used for monitoring MS treatment trials (i.e., new enhancing lesions and percentage T2 lesion volume changes) do behave in respect to clinical relapses in a way that Prentice suggested would be necessary for an outcome measure to be considered as a valid SE for the "true" clinical outcome of interest. The second study [7] shows that the occurrence of active T2 lesions and the changes in T2 lesion volume over the first year of the study account for slightly more than 50% of the difference in EDSS changes seen over the entire study duration in IFNβ-1b-treated patients when compared with those receiving placebo. This observation, if confirmed, would imply that, although IFNβ-1b was able – in this trial of SPMS patients, but not in others [8-10] – to reduce significantly the accumulation of clinical disability and T2 MRI abnormalities, the overall effect of IFNβ-1b on disability was not fully explained by its effect on the considered MRI metrics.

A meta-analytic assessment of the results of all, or nearly all, the randomized trials of MS which collected both clinical and MRI data would be a desirable and warranted approach to confirm these preliminary results. Clearly, the validation of MRI metrics for monitoring treatment effects in MS is a complex process, which would require an intensive collaboration and data-sharing between investigators and pharmaceutical companies, which hopefully should lead to a better definition of MRI surrogacy in MS.

References

1. Kurtzke JF (1983) Rating neurological impairment in multiple sclerosis: an expanded disability status scale (EDSS). Neurology 33:1444-1452
2. McFarland H, Barkhof F, Antel J, Miller DH (2002) The role of MRI as a surrogate outcome measure in MS. Mult Scler 8:40-51
3. Prentice RL (1989) Surrogate markers in clinical trials: definition and operational criteria. Stat Med 8:431-440
4. Sormani MP, Bruzzi P, Comi GC, Filippi M (2002) MRI metrics as surrogate markers for clinical relapse rate in relapsing remitting MS patients. Neurology 58:417-421
5. Sormani MP, Bruzzi P, Beckmann K et al (2003) MRI metrics as surrogate endpoints for EDSS progression in SPMS patients treated with IFNβ-1b. Neurology 60:1462-1466
6. Comi G, Filippi M, Wolinsky JS and the European/Canadian Glatiramer Acetate Study Group (2001) European/Canadian multicenter, double-blind, randomized, placebo-controlled study of the effects of glatiramer acetate on magnetic resonance imaging-measured disease activity and burden in patients with relapsing multiple sclerosis. Ann Neurol 49:290-297

7. Kappos L, Polman C, Pozzilli C et al (2001) Final analysis of the European multicenter trial on IFNβ-1b in secondary-progressive MS. Neurology 57:1969-1975
8. Li DKB, Zhao GJ, Paty DW and the University of British Columbia MS/MRI Analysis Research Group and the SPECTRIMS Study Group (2001) Randomized controlled trial of interferon-beta-1a in secondary progressive MS: MRI results. Neurology 56:1505-1513
9. Cohen JA, Citter GR, Goodman DA et al (2002) Benefit of interferon β-1a on MSFC progression in secondary progressive MS. Neurology 59:679-687
10. Panitch H, Miller A, Paty D, Weinshenker B for the North American Study Group on Interferon beta-1b in Secondary Progressive MS (2004) Interferon beta-1b in secondary progressive MS: results from a 3-year controlled study. Neurology 63:1788-1795

Chapter 11

Defining Responders and Non-responders

I. Aban, G. Cutter

Introduction

A physician in a clinic would typically like an answer to the question: Who will benefit from this treatment? In particular, will this patient benefit from this treatment? In an effort to answer the question, astute clinicians observe patients serially; in other words, they will try a treatment on the patient and then adjust the therapy if his or her condition does not respond. Although the sample size in this experiment is $n = 1$, the observations collected over time for this patient and others provide a knowledge base and have led clinicians toward a major understanding of certain treatments. Often, and historically, these understandings, taken collectively, have informed practice standards. However, this approach runs the risk of underestimating the extent of unsuccessful treatments, due to a failure to examine those patients who do not return.

In multiple sclerosis (MS), we measure our success in preventing progression of the disease when the patient has few relapses and suffers no increased disability, and also, more commonly, through the use of magnetic resonance imaging (MRI) parameters: for instance, no new lesions found by MRI, no increase in T2 volumes, and minimal brain atrophy. This system of evaluation differs from the response to symptomatic therapies, which focuses on symptom improvements such as: Did the rash go away? In the treatment of MS, we have not been able to follow this time-honored symptom management approach. Indeed, if this were done in MS, we would have to study the time to the disappearance instead of the incidence of gadolinium-enhancing (Gd) lesions.

In clinical trials, the response is often assessed by the effect of treatment; typically represented by the average treatment effect of the treated or control group. Did more patients fare better under treatment A, B, C, D, or the placebo? What characteristics of the patients are useful in *predicting* (with high sensitivity and specificity) who will respond to this treatment? In clinical practice, the response of interest is usually at the individual level: Is the patient relapse-free or is the patient free of MRI activity? Is the patient better? Furthermore, there may be no control group in clinical practice, and the meaning of "better" is often defined by the patients themselves.

How Does a Multiple Sclerosis Patient "Respond" to Treatment?

Some of the terms generally used to describe a clinical response are: non-response, poor response, partial response, breakthrough disease, and progression.

Are all these terms the same when studying MS? In infectious diseases, the absence of the organism or virus responsible for the infection is evidence of successful therapy or positive response. In MS, the absence of MRI activity or the absence of exacerbations, although necessary, are not sufficient reasons for us to say a person has responded. Thus, even the definition of responder is problematic in MS because our endpoints are not definitive.

Why is MS different? The condition of MS patients can worsen without exacerbations or MRI activity. Some patients have no brain MRI lesions and still do poorly from a clinical point of view. Early exacerbations or lesions can often be considered to be evidence of a delay in effectiveness or evidence of ineffective therapy. However, the absence of activity for periods of time is a hallmark of this disease. Progression or lack of progression on the Expanded Disability Status Scale (EDSS) may be too insensitive a measure to judge whether a clinical response has occurred. If a clinician uses the criterion "no exacerbations at any time" as the definition of a responder, it is likely that they will end up with no suitable treatment to offer the MS patient. Even the highly touted but now beleaguered Tysabri (natalizumab) has not been exacerbation-free! It was found to reduce exacerbations by two-thirds against a placebo; thus the longer the duration of observation, the more likely it is that a treatment will be seen to fail.

How do we define MRI responders? If new activity develops, the presumption is that this will translate into clinical worsening. However, the absence of brain MRI changes or activity is not, in and of itself, a guarantee of successful response. Our astute clinician will often make a within-person comparison, i.e., compare the current result with prior experience with this particular patient. Is this patient better functionally than before? Are there fewer MRI lesions than before? Is there less increase in MRI burden of disease? How do the composite scores for various items compare with those from previous visits? He/she may also use historical controls: Is this patient doing differently than his/her other patients? This approach or view of response presumes adequate historical information and recall of past experiences. These experiences are often biased by only seeing patients who return to the practice. Discontented patients often go elsewhere and, therefore, failures may be removed from the comparisons. In addition, we all have a tendency to remember our biggest successes and worst failures, with the remainder blending into the background, often making an accurate historical perspective difficult.

Responder Analysis

Responder analysis is the approach to understanding what works in whom. Responder analysis is not a new idea. In his book, Human experimentation: *A guided step into the unknown* [1], W. Silverman, quoting Galen, who was very influential and prophetic in medical research, says that "All who drink of this remedy recover in a short time, except those whom it does not help, who all die. Therefore, it is obvious that it fails only in incurable cases." Such personal recall and perspectives are, of course, not what we seek in our current attempts to identify responders, but this form of hubris may still lurk in our assessments.

In clinical medicine, while we work prospectively, it is often asked: What did I do for the patient and did it work? This is known as a retrospective approach. Quite naturally, clinicians doing research want to use the retrospective approach in identifying responders and non-responders and asking how they differ from each other. However, this is very risky and often leads to wrong conclusions. This difference between prospective and retrospective approaches can lead to serious disagreements between statisticians and clinicians over the best approach to adopt.

Cause and effect are key concepts in response analysis. A "cause-to-effect" must be defined a priori and is prospectively assessed, with the cause preceding the effect. On the other hand, a "cause-from-effect" retrospectively identifies differences by comparing responders and non-responders. Although convenient and easier to implement, the conclusions from retrospective studies may be misleading because of the confounding variables that were not controlled. These hidden or confounding variables may not be the predictors of interest, but may be related in such a way as to contribute to the effects observed in the study.

Responder analysis may be based on either observational study or clinical trials. **Observational studies** use data based on real patient populations and real standards of treatment. Because of the convenience that these studies offer to patients and physicians, observational studies are typically easier to conduct, have larger numbers of patients, and are more likely to have long-term observations. However, their main limitation is the lack of comparative controls. There will always be the presence of unknown confounding variables such as the decision-making process in the assignment of treatments. These processes are not random and often lead to selection biases that do not offer fair comparisons among treatments and also yield variable standards of treatment. This variability in how treatments are provided also causes variability in exposure to treatment. Due to the beliefs held by the study sponsor and authors who want to publish results, additional biases in assessments can occur. Often the consequences of these biases result in overestimating treatment effects and numbers of responders. Furthermore, when the characteristics of responders are examined, it can be the selection biases that are identified and not the actual predictors of response.

Clinical trials, on the other hand, are performed in a controlled experimental setting which minimizes the introduction of bias in the evaluation of the therapies. Treatment and assessment are well-defined prior to the collection of data; criteria for selecting participants into the study are defined in advance and lead to more homogeneity in the study population; and treatments are allocated randomly, which takes the selection biases of assignment out of the mix. Clinical trials usually involve a parallel matched control group which allows unbiased estimation of the response to therapy and prediction of responders separate from treatment effects. Some of the limitations of a clinical trial are the artificial selection of patients due to the restrictions placed by rigid entrance criteria. For example, participants may have the requirement for active disease and are identified based on explicit selection criteria. This results in a less generalizable group of patients, not only due to the more homogeneous population enrolled than that seen by the typical clinician, but also because of the effect of "tender loving care" experienced through being in a study, or a so-called "placebo effect." Being chosen to take part in a study may improve the patient's perception of their situation, making them "feel special," and this can be misinterpreted as a treatment effect. Also, those who volunteer in studies often differ markedly from those who do not volunteer. Finally, and one of the most controversial limitations, is that the random allocation of placebos may be perceived as being unethical.

In responder analysis it is, first and foremost, important to know the inclusion criteria used; in particular: What is the population from which the participants were selected and not selected, and were there any referral biases? One also needs to know how the treatments were allocated and how the patients obtained the treatment. Randomization is the gold standard when it comes to allocation of treatments because it is not influenced by any prior knowledge of the patient and theoretically creates treatment groups that are comparable in all respects, so that any differences after treatment can be attributed to the drug itself. Finally, it is critical that the outcome measures are clearly defined and carefully measured. This is the basis of determining who among the patients are responding. But what are the statistical tools used to describe what we mean when distinguishing between responders and non-responders, and how are the measurements made?

Risk versus Odds

In predicting responders, we want to know how much more likely is one group to respond to therapy relative to another group. Two commonly used measures to quantify this question are relative risk ratios (RR) and odds ratios (OR). Relative risk compares the incidence of two groups, i.e., RR = P_r/P_c where P_r is the incidence in the risk group and P_c is the incidence in the comparator or control group. This measure is usually the preferred comparison for assessing risk and/or predictors of response. The problem with relative risk is the need to obtain incidence figures. Incidence rates require two points in time; taking a population free of the condition but with a likelihood of it developing, and assessing the occurrence of new cases over the interim period. Obtaining such data requires prospective studies and is far more costly than examining odds ratios, which can be computed from cross-sectional data.

The odds ratio compares the odds of the risk group responding to a treatment relative to the odds in the comparison group, i.e., OR = O_r/O_c where $O_r = P_r/(1 - P_r)$ and $O_c = P_c/(1 - P_c)$. Therefore, O_r is the ratio of the probability of responding to not-responding to treatment in the treatment group, while O_c is the ratio of the probability of responding to not-responding to treatment in the control group. Although these two measures (RR and OR) both give estimates of the magnitude of association between the predictor and the response and are often used interchangeably by authors, they are not the same. The explicit mathematical relationship between these two ratios is given by the equation:

$$RR = \frac{OR}{(1 + P_c * (OR - 1))}$$

It should be clear from this equation that RR is smaller than OR because we are dividing the OR by a number greater than 1 to obtain the RR. Therefore, in general, the odds ratio overestimates the relative risk. However, when the probability of the response is lower than 10% or above 90% (i.e., response or non-response is a rare event), the odds ratio approximates the relative risk.

To illustrate, suppose that a MS study investigates the effect of two treatments on EDSS progression. In Table 1, A represents the number of patients on Treatment 1 (Rx 1) who worsened in spite of therapy and B those who did not. For patients on Rx 2, C represents those who worsened while D represents the number of patients not progressing (i.e., responding to Rx 2). Therefore, 70 out of 4800 patients worsened on Rx 1 and 50 out of 2400 patients worsened on Rx 2. The odds ratio of worsening is OR = (A/B)/(C/D) = (AD)/(BC) = 0.696, while the relative risk is RR = [A/(A+B)]/[C/(C+D)] = 0.70.

How do we interpret these ratios? Both ratios can have values between 0 and infinity. A value of 0 indicates no association between the treatment and response, while a value less (or greater) than 1 indicates a reduction (or increase) in the risk or odds of worsening under Rx 1 relative to Rx 2. Note that, in this example, the relative risk is approximately equal to the odds ratio because of the rarity of the response. The computed values indicate about a 30% decrease in the rate of worsening for patients in Rx 1 relative to patients in Rx 2.

Table 1. Results of a study with 7200 patients

	Worsen		
	Yes	No	Total
Rx 1	70 (A)	4730 (B)	4800 (A+B)
Rx 2	50 (C)	2350 (D)	2400 (C+D)
Total	120 (A+C)	7080 (B+D)	7200 (*n*)

These two measures are indicative of the ability to define responders and non-responders. That is, Rx 1 is more likely to produce a response, since it had fewer patients who worsened than with Rx 2. However, they are on a ratio scale which is often not intuitively understood. For example, a ratio scale implies that in magnitude, a relative risk of 0.50 has the same relative importance as a relative risk of 2.0.

While treatment is a basic variable that we all use in identifying responders and non-responders, it is not what we commonly think of when raising questions of responders. In general, we think of identifying responders on the basis of patient characteristics that are associated with better responses in one or both of the treatment arms.

An Illustration: Do EDSS and Age Matter in Gd Response?

Consider a study involving 200 patients who received treatment "X" with the outcome measure being the presence or absence of new gadolinium-enhancing (Gd) lesions. We are interested in investigating whether EDSS matters in Gd response; that is: Are patients with lower EDSS levels more likely to respond to therapy? Table 2a is a summary table that examines response based on whether the patients had an EDSS below 3 or an EDSS of 3 or higher. Based on this table, we can calculate the odds of responding by EDSS level. Here, the odds ratio is 3. This means that MS patients with an EDSS <3 have three times the odds of responding to the treatment than patients with EDSS ≥3. Similarly we could compute the relative risk of responding. Surprisingly, we see that the RR is equal to 2. In this situation, the incidence of responding is not rare and the odds ratio is overestimating the treatment benefit.

The fact that, in general, the odds ratio is not equal to the relative risk should be considered when discussing and/or analyzing data. Analyses based on logistic regression models estimate the odds ratio, while analyses that use the Cox proportional hazards regression model or Cox models are actually estimating the relative risk. Too often researchers interpret the results from logistic regression models to be relative risk and thereby often overestimate both treatment effects and/or the strength of predictors of response.

Effects of Incomplete Data or Dropouts

In the imperfect world of a clinical trial, patients sometimes drop out of the study. We again consider the example summarized in Table 2a. Table 2b adjusts the values in Table 2a to include some differential dropout from the study. Suppose that three of the cells experience a 20% dropout rate, but the cell for non-responders with EDSS ≥3 has a dropout rate of 40%. In this case, the actual data observed at follow-up are in Table 2b, not Table 2a. Here, the odds ratio is found to be 2.25, which is different from the odds ratio of 3 in the no-dropout case.

Table 2. Complete data (a) and data with dropouts (b)

(a)

Predictor	Responder	Non-responder	Total
AEDSS <3	50	50	100
EDSS ≥3	25	75	100
Total	75	125	200

(b)

Predictor	Responder	Non-responder	Total
EDSS <3	40 (80% of 50)	40 (80% of 50)	80
EDSS ≥3	20 (80% of 25)	45 (60% of 75)	65
Total	60	85	145

Effects of Age Considering EDSS

Next consider age in the analysis as another predictor of response variable. Of the same 200 patients from our example trial with no dropout, we now provide the breakdown of age and responders as shown in Table 3. We compute the odds ratio to assess the impact of age on being a responder. Here, the odds ratio is 2.5, which shows that age is an important predictor variable in response. Specifically, the younger patients are more likely to be responders.

Now, what happens if we simultaneously examine the relationship of age and EDSS on response? We subdivide the population into two 2 by 2 tables based on EDSS level and for each table we compare the responders by age. Using the values in Table 4a, b, the odds ratios when EDSS < 3 and EDSS ≥ 3 for the age variable are 1 and 8, respectively. We do not see the same impact of age on response

Table 3. Response to treatment versus age

Predictor	Responder	Non-responder	Total
Age <45	50	55	105
Age ≥45	25	70	95
Total	75	125	200

Table 4. Patients with EDSS <3 (a) and patients with EDSS ≥3 (b)

(a)

Predictor	Responder	Non-responder	Total
Age <45	30	30	60
Age ≥45	20	20	40
Total	50	50	100

(b)

Predictor	Responder	Non-responder	Total
Age <45	20	25	45
Age ≥45	5	50	55
Total	25	75	100

when we stratify by EDSS levels. These two odds ratios are very different, which implies that the relationship of age to the response depends on the EDSS status. This is known as an age-by-EDSS status interaction.

This illustrates the complexity of finding responders and considering response variables. Frequently the analyses end with univariate analyses, and if a predictor is found, success is declared. In the above example, there is no association between age and response for patients with EDSS <3, so declaring that age is a predictor of response is in general not true, although the data from Table 3 would suggest otherwise. Caution when declaring predictors of response is definitely needed.

Predicting Response

Sometimes predictors of response are illusory and stem from design criteria. For example, a common culprit is a phenomenon called regression to the mean. This design-induced response suggests that results are predicted by the baseline levels. When patients are selected on some criteria, such as active Gd lesions, then it is easily shown that the number of Gd lesions at baseline is a predictor of response. This is not because of some real phenomenon, but is merely the result of random error. Consider rolling a dice with the numbers 1 through 6 on each of the faces. Further, suppose the points on the dice (the numbers 1 through 6) represent the number of Gd lesions of that patient. Suppose we decide to use this dice as our screening tool and if a patient has a 4, 5, or 6, they are in the trial. Then if we were to screen 120 patients we would expect 60 of them to have a 4, 5, or 6.

What happens to our patients when we next see them? To keep this a purely artificial example, assume that we merely roll the dice again. The rolls of the dice are independent and for each of the 60 patients "enrolled" we would see 10 ones, 10 twos, etc. on the next roll of the dice. Thus, while the patients started the study with "lesions" of 4, 5, or 6, at the next visit they had values of 1, 2, 3, 4, 5, and 6. Since these are independent rolls of the dice, on average, a patient would have a value of 3.5, which is the average of 1, 2, 3, 4, 5, and 6. Table 5 shows the mean "reduction" in Gd lesions by baseline level. The correlation between the baseline score and change in Gd appears because the group with the highest baseline value is capable of the greatest fall.

Table 5. Hypothetical data for regression to the mean

Baseline Gd score	Number of patients at baseline	Expected follow-up mean	Change in Gd lesions
4	20	3.5	0.5
5	20	3.5	1.0
6	20	3.5	1.5

While many readers will quibble that this situation does not make biological sense, it can easily be shown that adding correlation, ρ, between the rolls of the dice reduces the appearance of this regression to the mean, but does not eliminate it. In fact, it can be shown that if R is the amount of regression toward the mean that would occur in the independent situation, $(1 - \rho)^*R$ is the regression toward the mean when the measurements at baseline and follow-up are correlated.

To consider our dice example a little further, if we have three dice and use the sum of the dice throws 1 and 2 as our screening values, and the sum of dice throws 2 and 3 as the follow-up value, intuitively it would seem that these total scores are correlated with a value of $\rho = 0.50$, since half of the sum in both instances comes from the same dice (dice number 2). This is correct mathematically as well. In such a situation, the regression toward the mean is reduced by the size of the correlation of what we would experience if we rolled four dice (two dice to represent screening and two more dice to represent follow-up).

But does this happen in reality? Figure 1 shows data from a trial conducted with an active disease-modifying therapy and a placebo. In both groups, we see an increasing reduction in the number of lesions, with an increase in the average reduction based on where we cut the baseline distribution. When we look only at individuals with any lesions, we see an average change of two lesions in the "placebo" group and four in the "treated" group. If we restrict our view to those with more than one lesion on entry, we see the placebo group drops almost four lesions and the treated drops almost seven. Examining those with more than two baseline lesions, we see that the placebo group decreases by slightly more than four and the actively treated group by almost nine. Thus, whilst it may be tempting to conclude that the baseline Gd counts are predictors of response, it is clear that caution should be exercised.

Finding a statistically significant association between a predictor and a response is not often sufficient in response analysis. Researchers would also like to predict the response. When this is the case, it is critical to assess the validity of the analyses, i.e., the ability to correctly categorize patients with pre-treatment characteristics as responders compared with patients without characteristics as non-responders. Specificity and sensitivity are the validity measures of most

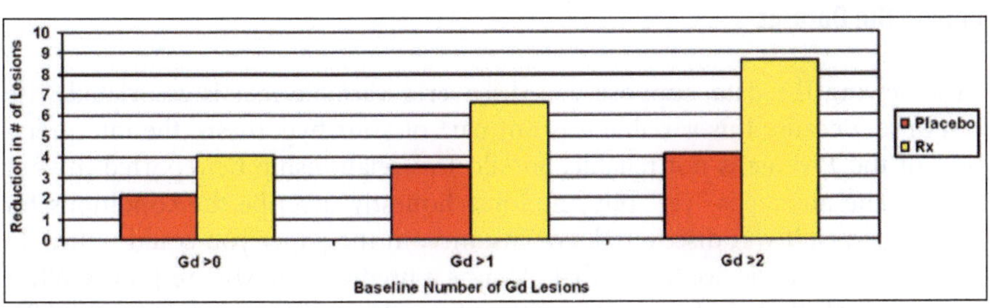

Fig. 1. Reduction in number of Gd lesions versus number of Gd lesions at baseline (*red* = placebo group, *yellow* = treated group)

Table 6. Response to treatment versus presence of Gd lesions

Predictor	Responder	Non-responder	Total
Gd–	50	50	100
Gd+	10	90	100
Total	60	140	200

interest from a statistical point of view, while positive predictive value (PPV) and negative predictive value (NPV) are the validity measures of most interest from a clinical point of view.

Sensitivity is defined as the probability of correctly predicting that a person is a responder given that the person actually responds to treatment. Specificity is the probability of correctly predicting a non-responder given that a person will not respond to the treatment. Positive predictive value provides a measure of how often we are correct in predicting responders among those for whom we expect to respond. If we predict a non-response (or a negative result), we use the term negative predictive value to provide a measure of how often we are correct in predicting non-responders. To illustrate this, if the presence of Gd lesions is associated with a poor response, we can use the values shown in Table 6 and get sensitivity = 50/60 = 83%, specificity = 90/140 = 64%, PPV= 50/100 = 50%, and NPV= 90/100 = 90%.

The PPV and specificity are not really very good, since only half of the cases predicted to have a good response in fact do. Another measure, the Number Needed to Diagnose (NND) or predict a case, is used to assess just how good the values shown above actually are. This number is defined as: NND = 1/[Sensitivity – (1 – Specificity)]. Thus, in the above example NND = 1/[0.83 – (1 – 0.64)] = 1/(0.83 – 0.36) = 1/.47 = 2.13. This implies that for approximately every two patients predicted to respond, only one actually responds. Many good diagnostic tools have a value of NND close to 1.

Other Issues in Responder Studies

Mining the Dataset

In reviewing the data, suppose you discover a variable that is associated with response. Because this variable was not part of your hypothesis, the interpretation of the *P*-value is not reliable. Should this relationship be reported in your paper? The answer is "yes" but one must honestly describe the conditions for which the result was discovered. For instance, in the paper you could write: "In exploratory analysis, we found that "X" was a predictor of MRI response. While the nominal *P*-value for assessing the strength of this association is 0.001, because of the exploratory nature of the analysis, we encourage caution in the interpretation of this *P*-value and encourage replication of the findings." This is translated to mean that we were searching our data for predictors of response. We want to be on record as the being first to report this, but because we were just poking around when we found the relationship it would actually be misleading. We really hope that you see this relationship in your data too.

Search for Predictors of Response

Suppose I am interested in doing a study which examines the relationship of gender and smoking to MRI response in patients treated with interferon (IFN). I perform an analysis in the IFN-treated arm with 834 patients where 656 (79%) are female and 178 (21%) are male. My hypothesis is that the difference between females and males in the time until new MRI activity is seen is greater for smokers than non-smokers. An appropriate approach would be to do an overall analysis of time-to-event (the event being MRI activity) by gender followed by a subgroup analysis of gender among smokers and non-smokers.

The preliminary findings show that there is a 43% increase in new MRI activity among females as compared to males, and this increase is not significant. However, what I am really interested in is the difference between males and females in strata defined by smoking status. The conclusions, as presented in my paper, would be that there is a trend for females to have higher new MRI activity. This effect is primarily found among smokers, where the risk is nearly three times that of males. This confirms my belief that gender is important and underscores the importance of smoking cessation in females.

What really happened in the preparation of this paper? First, we did an analysis of the history of optic neuritis. We found that the overall test was not significant, i.e., none of the predictor variables in the model was found to be significant. We then looked for significance in subgroups of these variables: cigarette smoking (2 groups), age (3), sex (2), race (3), previous exacerbations (4), alcohol intake (2), BMI (3), previous IFN treatment (2), randomization date (3), family history (2), MRI activity at baseline (4), and gender (2). The total number of subgroup analyses performed taking each variable separately is 32, of which none were significant. We continued our subgroup analysis considering interactions of these variables. Although gender was not significant overall, as shown in Figure 2, we found a significant association in smokers (see Fig. 3). Then we remembered an article in *Neurology* by Riise et al. [2] entitled "Smoking is a risk factor for multiple sclerosis". Therefore, after numerous analyses done, we finally found an interesting result that we would be able to report which confirmed our biases and sense of the literature. Was this cheating? If so, how badly did we cheat? The significant result about smoking is publishable, but with unknown value. This is because this wasn't the plan of the study, but an afterthought decided upon after reading some of the literature. As seen in Figure 4, the probability of observing at least one significant association under the null hypothesis of no association increases as the number of tests increases.

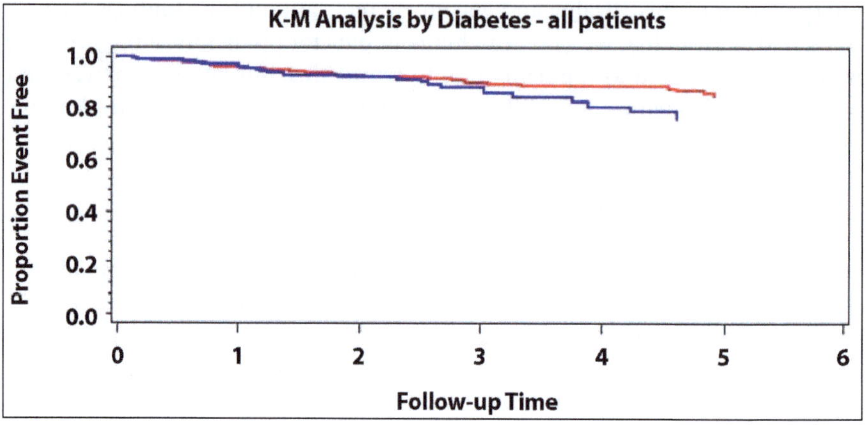

Fig. 2. Time to MRI activity by gender (*red* = male; *blue* = female)

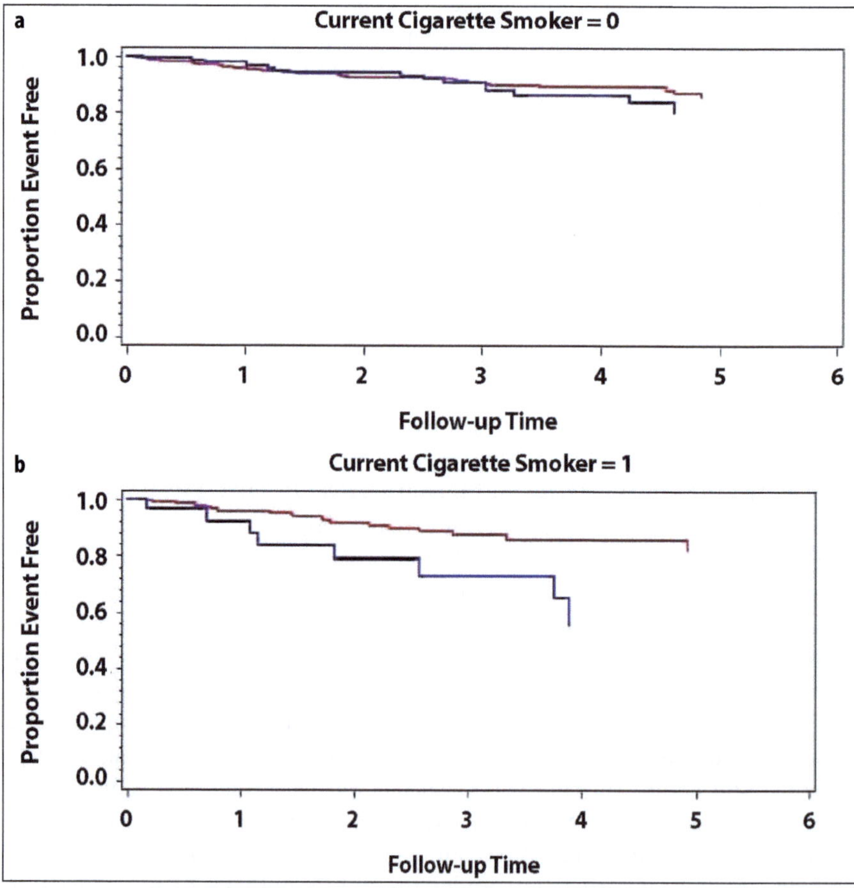

Fig. 3a, b. Time to MRI activity by gender stratified by smoking status; **a**, non-smokers, **b**, smokers (*red* = male; *blue* = female)

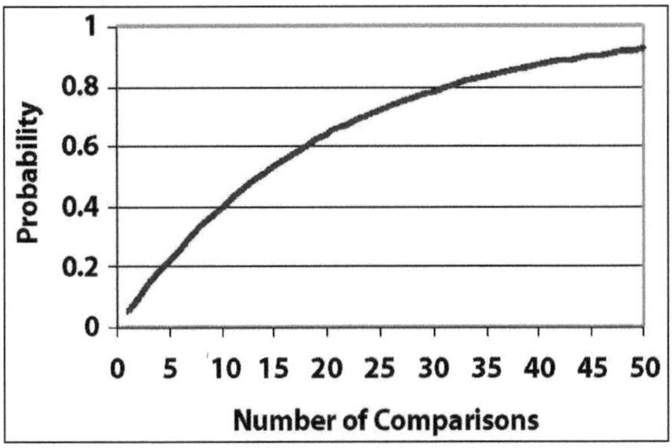

Fig. 4. Probability of at least one significant association compared with number of comparisons

In conclusion, defining in advance the question one wishes to answer is critical in response analyses. One must plan ahead and consider dropouts and losses. Predictor variables must be defined a priori in order to minimize biases. Bias is the "invisible" demon. Correlation, large odds ratios, and small *P*-values do not necessarily imply that the predictor variable of interest causes the response. Finally, one should be honest in the report and must not be blind to one's own biases.

References

1. Silverman W (1986) Human experimentation: a guided step into the unknown. Oxford University Press, New York
2. Riise T, Nortvedt MW, Ascherio A (2003) Smoking is a risk factor for multiple sclerosis. Neurology 61(8):1122–1124

Chapter 12

Predictive Models in Multimodal Imaging

K. Mouridsen, L. Østergaard

Introduction

The disease mechanism of multiple sclerosis (MS) causes progressive subcellular and cellular changes that may ultimately be detected by magnetic resonance imaging (MRI): for instance, in normal-appearing white matter (NAWM) the effects of the disease gradually alter the macromolecular and cellular compartmentalization of water, causing subtle changes in magnetization transfer ratio (MTR) and diffusion-weighted imaging (DWI). Similarly, MS lesions are characterized by serial image changes in several MR image modalities.

Early diagnosis, as well as rapid, cost-effective development of novel therapies for MS treatment, depend critically on sensitive, early markers of the disease and models of disease progression. MRI, having demonstrated characteristic changes at many disease stages, is ideally positioned for this purpose. The challenge remains, however, to integrate and interpret large data sets of multimodal, high-resolution MRI data in order to most efficiently detect changes that are characteristic of the disease, its evolution, and its successful treatment. Examples of the progress already made in this respect include the current analysis of MTR and DWI changes in NAWM: rather than using global averages of signal intensities as a metric of disease, the wealth of information from individual image voxels is utilized in histogram analysis, providing more insight and sensitivity to characterize disease progression. The aim of this overview is to extend these efforts, providing a general framework for exploiting the full statistical power of multimodal, repeated measurements in *single-voxel* images to characterize disease and disease progression.

This chapter reviews and discusses how to fully utilize the power of typical, state-of-the-art multimodal, longitudinal studies of NAWM or T2 lesion evolution in MS. We begin by discussing very general image comparison techniques, and then devote the rest of the chapter to the description of techniques for modeling and analyzing disease effects at the level of single voxels. We first present methods for detecting temporal changes – both local and diffuse – in images acquired at two time points. This is followed by the presentation of a general framework for predicting general outcomes such as evolution of T2 in a specific area, reversal of a T2 lesion evolution, or long-term neurological deterioration, using multiple MR modalities. After a discussion of typical analysis objectives such as variable selection, collinearity, and assessment of the relative importance of imaging predictors, we briefly discuss extensions to longitudinal studies. Finally, we discuss how predictive models may be used to evaluate treatment efficacy.

In our descriptions, we have emphasized the techniques implemented in well-established statistical packages, especially S-PLUS and SAS (see the Note on Software section), and provide references to more exhaustive details of the techniques as well as practical recommendations with respect to choice of test options. Throughout this chapter, MT and T2 hyperintensities will often be mentioned as target images, but this is for the purpose of illustration only. The techniques apply to any modality upon appropriate standardization/normalization to allow comparison of images. Many of the statistical techniques utilized in this review represent very modern approaches in this specific context. As these are not typically covered by reference textbooks, central references are provided throughout the chapter.

Although we begin this review with a discussion of image comparison techniques requiring no prior co-registration, we will throughout this chapter emphasize the statistical power and modeling flexibility gained with co-registration and subsequent measurement of focal (isolated) effects at the level of individual voxels – effects that are otherwise masked in the histogram analysis. The full details regarding co-registration of serial data in cases of progressive atrophy are beyond the scope of this chapter. For co-registration methodology and software, see Collins et al. [1] and the Montreal Neurological Institute (MNI) software (http://www.bic.mni.mcgill.ca/software/) or Fishl et al. [2] and the FreeSurfer software (http://surfer.nmr.mgh.harvard.edu/).

Analysis of Non-Coregistered Images

In this section, we discuss summary statistics of images across time points or subjects with no prior co-registration.

One of the fundamental experiments in medical imaging is concerned with characterizing the differences in a single image modality between two groups of subjects. The simplest approach for detecting a difference is to characterize each subject scan by a *single* number such as the average intensity over the imaging slice or volume, and then compare these means using the two-sample t-test (if the subject means are normally distributed) or the Wilcoxon (Mann-Whitney) test. However, reduction of the original data, such that the number of observations equals the number of subjects, prompts the need for a large number of subjects in order to obtain enough power to detect subtle and focal differences. Moreover, there is a risk that focal increases and decreases in image intensities cancel out when intensities are averaged over a large volume.

It has been suggested that properties of the intensity distributions, represented by histograms, may serve as a basis for a more detailed investigation of group differences. This way each subject is characterized by multiple distributional indices instead of only the intensity mean value. Widely used measures are histogram peak height, quartiles, and location of peak height [3]. Hence each subject is characterized by k values $x_1\ldots, x_k$, and we wish to test whether the mean value of the first measure, x_1, is the same for the two groups, and so on for $x_2\ldots, x_k$. It is not recommended that the standard t-test should be performed for each of the k measures, since this will result in a multiple comparisons problem and the combined effect of small differences will not be detected. Therefore, a multivariate test for equal means in the two groups across the k distribution characteristics such as the Hotelling T^2 test [4] should be performed. If the result of this test is significant, then the groups differ with respect to the measured distribution characteristics.

An alternative strategy for simultaneously testing if summary measures differ between two groups is to use $x_1\ldots, x_k$ as predictors (regressors) in a *logistic regression* with group as outcome. Hence, we use the distribution characteristics (or other subject-specific measures) to predict which group a subject belongs to. If a test of no regression is nonsignificant (all coefficients are zero), this indicates that none of the studied measures differ between groups [5]. In contrast to Hotelling's test, inspection of the fitted logistic regression model reveals which particular characteristics differ between groups, and hence we do not have to resort to univariate tests which ignore the combined effects of factors. Furthermore, the logistic regression model does not require any distributional assumptions and can be used even when the k subject characteristics are a mixture of continuous (e.g., mean image intensity) and categorical (e.g., patient subgroup) measures (see the section on Predictive Models for more details on logistic regression and multivariate modeling).

There are a number of other approaches to comparison of distributions including the Kolmogorov-Smirnov test, the chi-squared test for equal bin counts, and methods based on kernel density estimates [6]. For the purpose of detecting differences in MR images, a major drawback of histogram-based methods is that focal changes (involving a small proportion of the total brain pixel number) are not easily noticed, due to their negligible impact on histograms comprising thousands of voxels. As discussed below, reducing the huge number of observations to a few summary parameters represents a less efficient use of the available data, severely reducing the potential power of tests. Finally, differences in histogram summaries are difficult to interpret, as infinite combinations of (i) the number of pixels (tissue volume) showing image intensity change and (ii) the extent of individual pixel value change may result in a given histogram change. In addition, compromising the biological interpretation of results, this "convolved" biological heterogeneity poses a problem in the power analysis of studies estimating how many subjects are required.

Detection of Temporal Changes

In the following sections, we discuss the detection of group differences in temporal evolutions based on co-registered data.

Characterizing Change in a Single Subject

Graphical Techniques

Consider the case where subjects are scanned at two time points to assess changes in, for instance, MTR. To obtain an initial impression of the change in MTR for a single subject at the level of individual voxels, the voxel MTR values at time point t_1 can be plotted against corresponding values at time point t_2. The identity line should be added to the plot, since this represents the case where no change in MTR has occurred. Any deviations from this situation are of interest, and these are most easily identified when the identity line is plotted as a reference. Fitting a regression line is only of secondary interest in the visual display. Additional graphical techniques, which are useful for evaluating the relation between two measures, have been presented in other research [7, 8].

Linear Regression

In this section, we present methods for characterizing temporal changes in an image modality for a single subject, and discuss formulations of appropriate hypotheses and tests for detecting clinically relevant alterations.

If we let x_i denote voxel intensities observed at the first time point and let y_i represent the value in the same voxel at the second time point ($i = 1,2,\ldots,n$, where n is the number of voxels), then we can use the standard linear regression model to describe the change in voxel values over time. The linear regression model has the form:

$$y_i = \alpha + \beta x_i + \varepsilon_i$$

where ε_i has a zero-mean Gaussian distribution with variance σ^2. The hypothesis of no change in MTR for a single subject corresponds to testing $H_0 : \beta = 1$ against the alternative $H_0^A : \beta \neq 1$. We assume, for simplicity, that the intercept α is without interest, but all tests described in this section apply to both parameters, and we only focus on the slope, β, for simplicity. Note that standard statistical packages test $\tilde{H}_0 : \beta = 0$ against $H_0^A : \beta \neq 0$. However, the significance of H_0 can easily be assessed using confidence intervals, which are usually supplied by statistical computer programs. Essentially, the 95% confidence interval for β contains all the values of this parameter which would not be rejected in a corresponding hypothesis test at the 5% level. Hence H_0 should be rejected if the 95% confidence interval for β does not contain the value $\beta = 1$. Note that the hypothesis test described in this and the following sections rely on the assumption that the error term, ε, is normally distributed. Therefore this assumption should be verified whenever possible.

It is important to realize that, in many situations, point hypotheses like $H_0 : \beta = 1$ (or $H_0 : \beta = 0$) are inappropriate. In the planning of experiments, we are often interested in how many subjects should be included in a study in order to obtain a certain power for a specific test. More precisely: the probability of correctly detecting a deviation of a given size from the null hypothesis increases with the number of observations. Conversely, if we have enough observations, any effect of β away from H_0, no matter how negligible, will be judged significant. This poses a serious risk of overinterpreting trivial results, especially in the field of image analysis where the number of observations, in the form of voxels, is large. One way to minimize such a risk is to distinguish between statistical and clinical significance. We may say that if the value of β deviates from $\beta = 1$ by less than some small δ, then this difference is of no clinical interest (see Fig. 1).

Hence a more appropriate hypothesis is:

$$H_0 : | \beta - 1 | \leq \delta \text{ against } H_0^A : | \beta - 1 | > \delta$$

If we reject H_0, we may infer that there is a substantial decrease (or increase, since this is not a directional hypothesis) in MTR. Such tests are much more informative than the common point hypotheses and can be constructed from standard tests by using the union-intersection principle [4].

It is important to realize that the appropriate null hypothesis depends on the nature of the question to be examined. When we test a hypothesis, we can control the level of significance, which is usually set to 5%. Hence we are able to control the probability of rejecting the null hypothesis, when the null hypothesis is true. This is also referred to as the probability of a Type I error. In the setting above, there is a 5% risk of declaring a change in MTR (rejecting H_0), when no change is actually occurring (H_0 is true). However, if the model above is used to screen patients for MS, it is important to control the risk of (falsely) declaring no change in MTR when in fact MTR is declining. In this situation, the appropriate hypothesis is:

$$H_0 : | \beta - 1 | > \delta \text{ against } H_0^A : | \beta - 1 | \leq \delta.$$

Here we hypothesize that the regression line is not the identity line (i.e., we assume MTR changes). If this hypothesis is rejected, we may infer that the value of the slope is 1, meaning no change in MTR. With this hypothesis, we can control the risk of falsely rejecting the null hypothesis (declaring no change in MTR) when the null hypothesis is indeed true, simply by adjusting the significance lev-

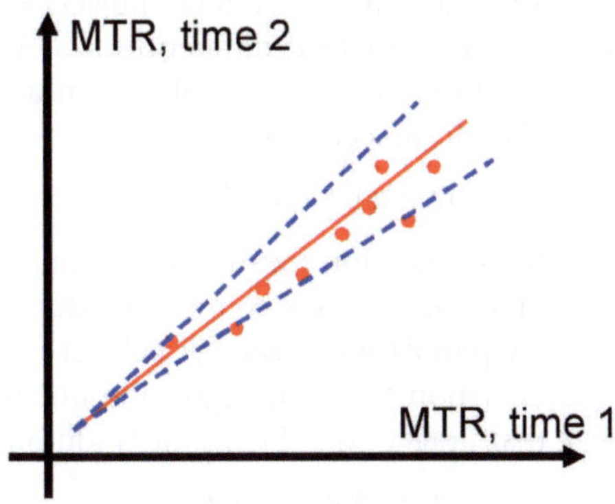

Fig. 1. MTR values observed at two time points. Changes in MTR are only considered clinically relevant if the fitted regression line deviates from unity by more than a pre-specified value, δ

el of the test. Effectively, we therefore control the risk of failing to detect actual MTR changes which may be indicative of MS.

The hypothesis tests described above rely on distributional results which require that observations are independent. To ensure independence, not all voxels in an image should be included in the calculations. Instead voxels should be sampled such that neighboring voxels do not appear in the model. This will ensure that the spatial correlation in images does not affect the hypothesis tests.

Hierarchical Models

Having described the analysis of single subject changes, we now present methodology for characterizing developments in a group of subjects. In the following section, we show how to compare changes between groups of patients.

Multiple Subjects

Suppose n subjects have been scanned at two time points, and we wish to characterize the developments in MTR in this group. Then we may fit a single linear regression model to the pooled data to get an estimate of overall slope and intercept. However, ignoring the subject effect, the residuals for individual subjects may have non-zero mean (see Fig. 2). This means that the fitted model systematically over- or underestimates responses for some subjects and results in a large residual standard error.

To minimize this subject effect, we can fit regression models to each subject, obtaining estimates α_i and β_i for $i = 1,\ldots,n$, where n is the number of subjects. As can be seen in Figure 2, the residuals in this model are much smaller in magnitude and centered around zero. In this example, the residual standard error (σ) is about one-seventh of the standard error in the single-slope model.

But there are serious problems with this model, although we have accounted for the subject effect. First, as we allow separate slopes for each subject, the number of parameters in the model increases linearly with the number of subjects; but more critically, by modeling each subject individually we are unable to make inferences about the population from which these subjects were sampled. Hence, in our final model we will assume that the differences in slopes between individuals are due to random variations around a population mean. More formally, for the jth measurement on the ith subject we have:

$$y_{ij} = (\alpha + a_i) + (\beta + b_i)\, x_{ij} + \varepsilon_{ij} \qquad (1)$$

Here α and β represent the population mean intercept and slope (fixed effects) for this group. The variables a_i and b_i are zero mean random variables representing the deviation from the population mean for subject i (random effects). As usual, ε_{ij} represents the deviation from the regression line for the jth measurement on subject i. Note that in fact Eqn (1) has the traditional regression form:

$$y_{ij} = A + Bx_{ij} + \varepsilon_{ij}$$

with the modification that $A \sim N(\alpha, \sigma_a^2)$ and $B \sim N(\beta, \sigma_b^2)$. The standard deviations of a and b are estimates of the between-subject variability.

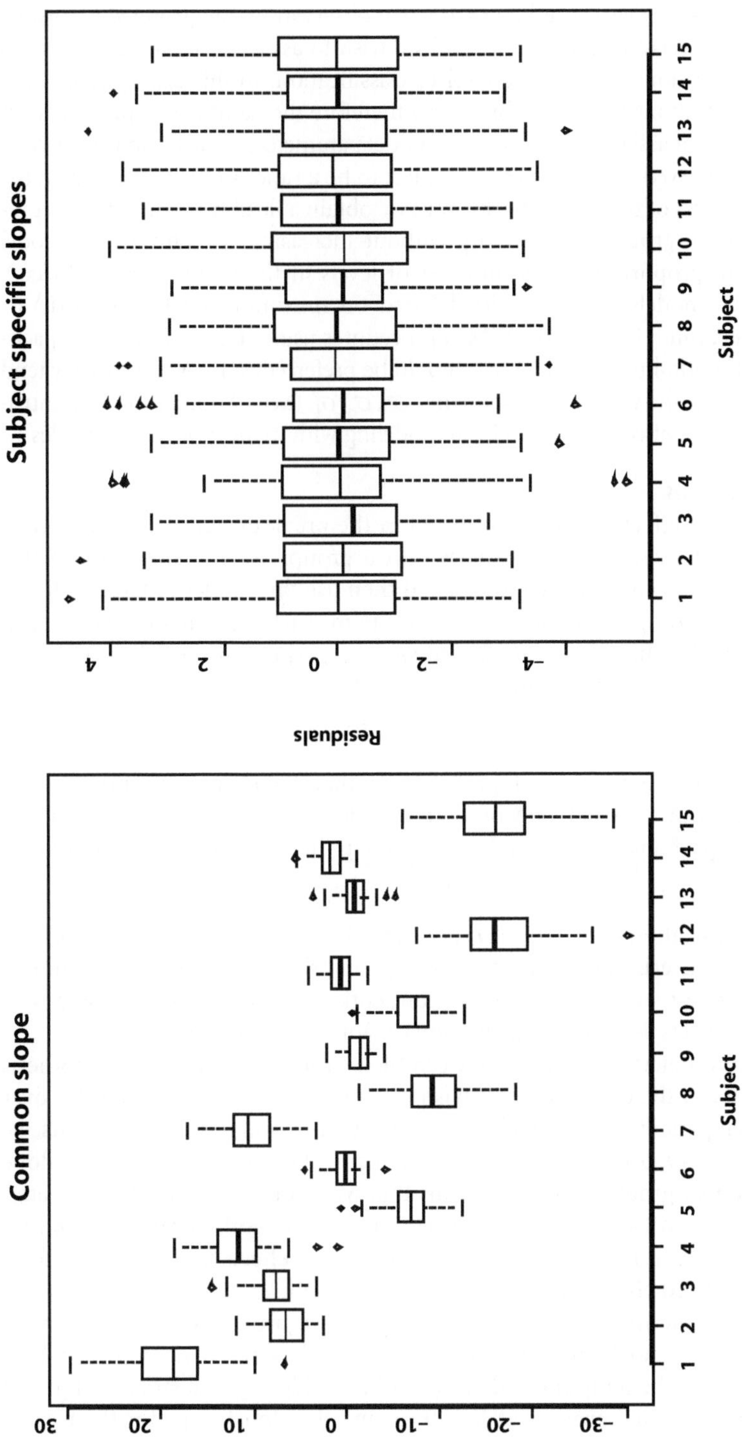

Fig. 2. Box-plots of residuals for each subject in a linear regression model with a global slope (*left*) and a model where a slope is fitted for each subject (*right*). Note that when fitting a single, global slope, the subject residuals have non-zero mean value. When individual slopes are fitted, the residuals for each subject are much smaller and have mean values close to zero

Mixed-effects models should be used when observations are grouped according to one or more classification factors and we wish to associate a common random effect to observations at the same level of classification. In this example, we have multiple observations for each subject, and therefore the natural grouping factor is subject. Instead of assigning one or more model parameters to each subject, we consider the unobserved parameters for a subject to be a random fluctuation from the global parameters of the group. In this way, we obtain a flexible model which represents the within-group heterogeneity without increasing the number of model parameters in proportion to the number of levels in the group (here: subjects). Mixed-effects models are easily fitted based on maximum likelihood (ML) or restricted maximum likelihood (REML) procedures using standard statistical packages. For estimation purposes, REML should be preferred, since ML estimates tend to underestimate the variance components, σ^2, σ_a^2, σ_b^2. (See [9] for a thorough introduction to mixed-effects or hierarchical modeling with many practical examples.)

Group Comparisons

With the mixed-effects model introduced in the previous paragraph, we have a natural characterization of subjects within a group. We can now extend this model to include two or more groups, and then use this model to test for differences between groups. If we let x_{ijk} denote the measurement at time point t_1 in voxel i in the jth subject in group k, and we let y_{ijk} represent the follow-up measurement at time t_2, then we can write:

$$y_{ijk} = (\alpha_k + a_i) + (\beta_k + b_i)\, x_{ijk} + \varepsilon_{ijk} \tag{2}$$

This expression is equivalent to Eqn (1), except that the mean slope and intercept has an index $k \geq 2$, which means that we allow these parameters to change between groups. Equation (2) can be rewritten in the traditional regression form:

$$y_{ijk} = A_k + B_k x_{ijk} + \varepsilon_{ijk},$$

with the modification that $A_k \sim N(\alpha_k, \sigma_a^2)$ and $A_k \sim N(\beta_k, \sigma_b^2)$. Hence each group is initially characterized by a unique intercept and slope, but we explicitly model the heterogeneity within a group by representing the intercepts and slopes of individual subjects as random fluctuations around the group mean.

We can now detect differences between the k groups by testing the hypothesis $\beta_1 = \ldots, = \beta_k$. In the case where we wish to compare the decrease in MTR over time between patients and controls or between treated and untreated patients, we have $k = 2$. If the test for $\beta_1 = \beta_2$ is significant, there is a difference in slope between the two groups, and this means that MTR decreases at different rates. Most statistical software has procedures for tests of both fixed and random effects. In general, it is recommended to test random effects using the likelihood ratio test and fixed effects with conditional F-tests [9].

To illustrate the mixed-effects group analysis, we present simulated data for patients receiving no treatment (Fig. 3) and patients receiving a drug which potentially curbs the temporal reduction in MTR (Fig. 4). The simulated global slope for the untreated group was $\beta_1 = 0.65$, and the global slope for the treated

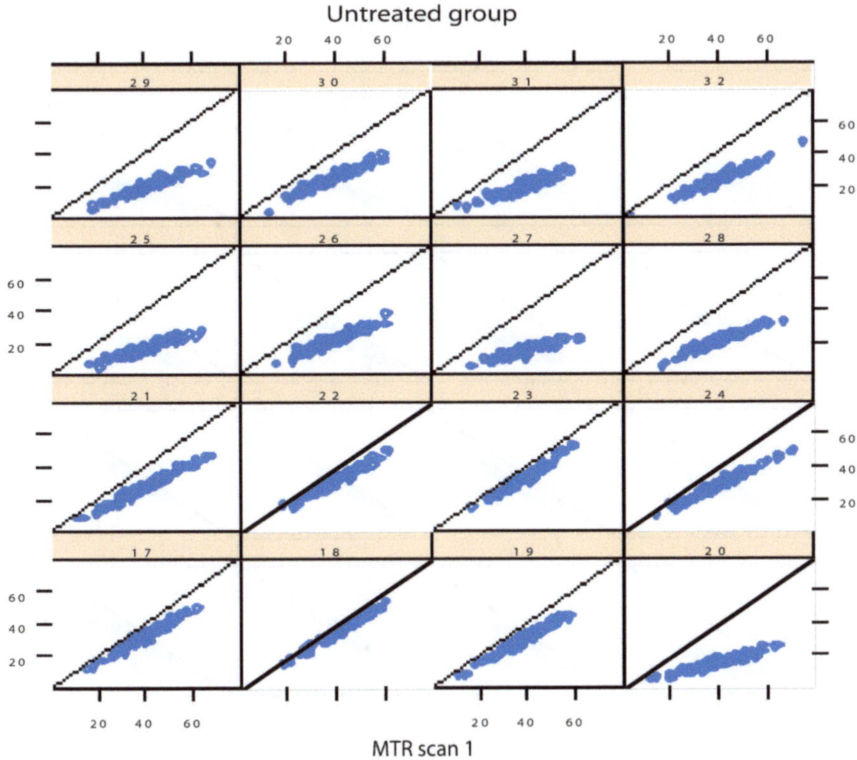

Fig. 3. Simulated data representing MTR values observed at two time points for patients receiving no treatment

group was set to $\beta_1 = 0.80$. Slopes for individual subjects were generated from Gaussian distributions with means β_1 and β_2 and standard deviations $\sigma = 0.15$ to simulate within-group heterogeneity.

By fitting the mixed-effects model, Eqn (2), using the NLME package in S-PLUS, we obtained estimates $\hat{\beta}_1 = 0.63$ ($\pm$ 0.06) and $\hat{\beta}_2 = 0.76$ ($\pm$ 0.04) of the global group slopes. The within-group standard deviation in slope was 0.16, which is very close to the simulated value of 0.15. The conditional F-test for equal slopes in the two groups gives $P = 0.017$, which shows that the MTR evolution is significantly different for treated and untreated patients. The likelihood ratio test for no random effects has a P value less than 0.0001, indicating that accounting for the within-group random effects leads to a significantly better description of the data.

Anatomical Mapping of Local Changes

This section provides an overview of techniques that yield high sensitivity in detecting subtle changes in image intensity for voxels or areas over time – suggesting disease progression.

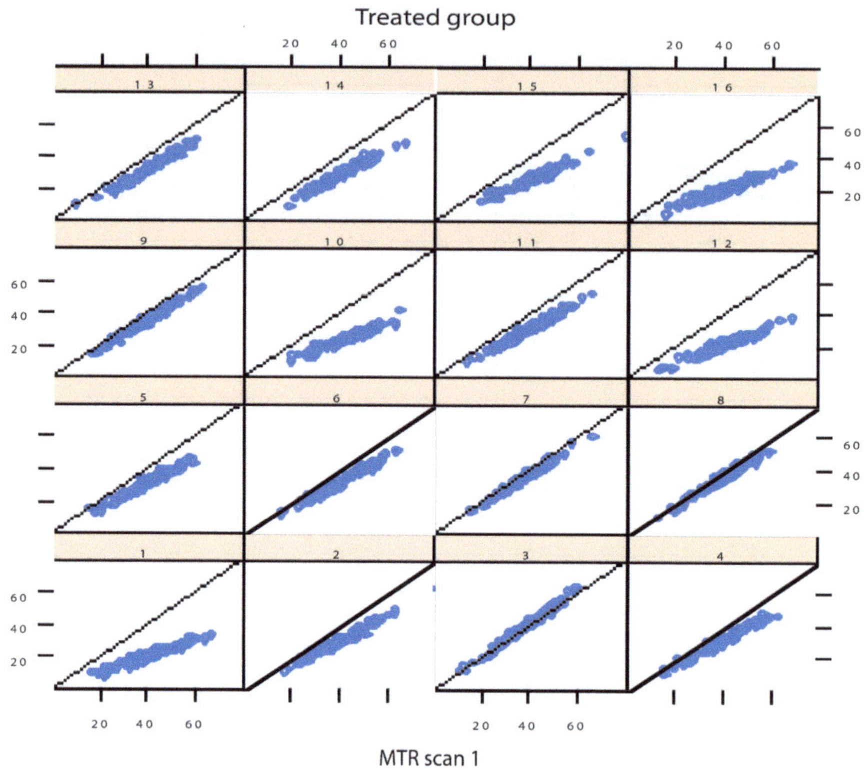

Fig. 4. Simulated data representing MTR values observed at two time points for patients receiving a treatment which is hypothesized to slow down the decline in MTR values

Physiological and biochemical alterations in brain tissue associated with a disease process may not be uniformly distributed within the brain. Moreover, changes in some regions may be subtle [10], and may therefore be masked in analyses of histogram parameters or regression, where the intensities of all voxels are pooled. To be able to detect subtle regional changes with reasonable power, it is necessary to effectively carry out a group comparison for each tissue voxel. Approaches for detecting local differences in brain images have been developed primarily within the fields of positron emission tomography (PET) and functional MRI (fMRI), and software utilizing the appropriate computational algorithms is now widely available. (For recent applications in MS see, e.g., [10, 11]).

As in the previous sections, assume we acquired images from n patients scanned at two time points. We were interested in detecting local changes in, e.g., MTR. Considering a single voxel location, we then have values $x_1, \ldots, x_n$ for the n subjects at time point 1 and $y_1, \ldots, y_n$ at time point 2. Hence the usual t-statistic could be calculated for this voxel to obtain an estimate of the local change relative to the standard deviation:

$$t = \frac{\bar{d}}{\frac{1}{n-1}\Sigma_{i=1}^{n}(d_i - \bar{d})^2}$$

where d_i, $= x_i - y_i$ and $\bar{d} = 1/n\ \Sigma_{i=1}^{n}\ d_i$. P values can be calculated as usual since this t-statistic has a t distribution with n degrees of freedom (see Fig. 5). However, since the number of subjects is typically small, this test is not very powerful for detecting local change because the number of degrees of freedom is low. Therefore strategies to increase the number of degrees of freedom should be pursued. In the extreme, the standard deviations in the individual voxels could be averaged to obtain a global estimate of the standard deviation. However, the voxel standard deviations typically depend on the observed differences d_i and exhibit systematic variation complying with anatomical structures. This dependence can be assessed simply by inspecting the map of local standard deviations.

There are no hard and fast rules for how to optimally increase the degrees of freedom in the t-statistic. Specific procedures depend on the extent and structure of the image noise. In PET images, the anatomical variation is often negligible and therefore a pooled, or locally pooled, estimate can be employed. In fMRI studies, where subjects are scanned under several conditions (relative to the two conditions described in this paragraph), the effective degrees of freedom can be increased by pooling the voxel standard deviations over these conditions. We refer the reader to the programs FMRISTAT (http://www.math.mcgill.ca/keith/fmristat/) developed by Keith Worsley, and SPM (http://www.fil.ion.ucl.ac.uk/spm/), initially developed by Karl Friston. The documentation and papers accompanying these programs discuss the issues of standard deviations in detail. The intricacies of computing voxel P values based on the t-statistics, when these are not independent, are discussed by Worsley et al. [12].

The analyses in this section can be extended to comparisons of temporal evolution between groups of patients or patients and controls. In this case, the two-sample t-test should be calculated in each voxel based on the subtracted images in the two groups (see Fig. 5). In general, using the theory of Worsley [13], maps of P values, corrected for multiple comparisons, may be calculated for voxel-wise t, F, and χ^2 tests.

Predictive Models in Multimodal Imaging

Many MR modalities have been introduced for diagnosis and monitoring of MS [14]. Each of these imaging techniques is used to identify isolated tissue characteristics associated with the disease. While much can be learned about the pathophysiology of MS through analysis of single modalities, mathematical modeling of several modalities and/or time points allow us to model and predict disease progression, and indeed evaluate drug efficacy by comparisons of predicted disease progression under different treatment strategies. This is done by combining the individual power of several modalities into models of the disease progression. As described below,

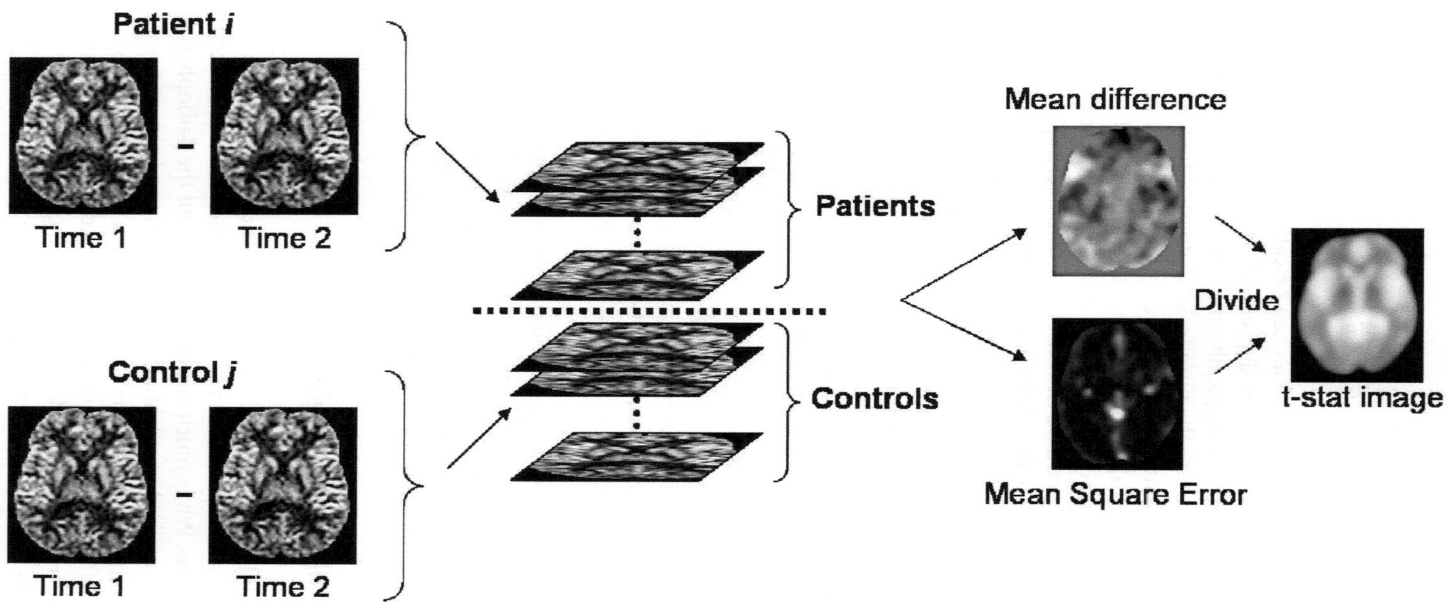

Fig. 5. Computation of an image of *t*-statistics, in a case where differences in an image modality observed at two time points are compared between patients and controls. The two acquired images for each subject are subtracted, and the subtraction images are averaged in the two groups. Then a *t*-statistic can be calculated in each voxel by dividing the average group difference by an estimate of the local standard deviation

this model allows separate modeling of the significance and relative importance of various imaging modalities in detecting disease-specific changes in MS.

Multivariate analysis is particularly important for assessment of the strength of individual MR modalities and prediction of disease progression, because the effects of certain parameters may depend on other modalities. In univariate screening of individual disease markers, the significance of single modalities can be masked due to lack of control over confounding factors [15]. This may lead to false rejection of important predictors of disease progression. Critically, we do not recommend selecting disease markers for use in subsequent multivariate analysis based on significance measures in univariate analysis. Instead, all relevant variables should initially be entered into a multivariate model. Methodology for assessing the significance of individual predictors in multivariate models is discussed below.

The type of multivariate modeling to employ depends on the specific endpoint(s) of the study. In many experiments, the endpoint is categorical. For example, we may be interested in whether a T2 lesion will develop based on changes in, e.g., MTR [16-18] or DWI metrics [19], or whether the Expanded Disability Status Scale (EDSS) will increase to above a certain threshold within a fixed time period [20, 21]. This type of modeling has been applied to studies of other (neurological) diseases, particularly stroke, where the endpoint is defined for each tissue voxel as infarction or no infarction [22, 23]. For such binary outcomes, the most natural model is logistic regression, which specifically models the probability of each outcome. In acute stroke, using the combined predictive strength of both perfusion and diffusion images, one obtains a probability map, where the risk of infarction is evaluated for each tissue voxel (see Fig. 6). These maps are useful for comparing

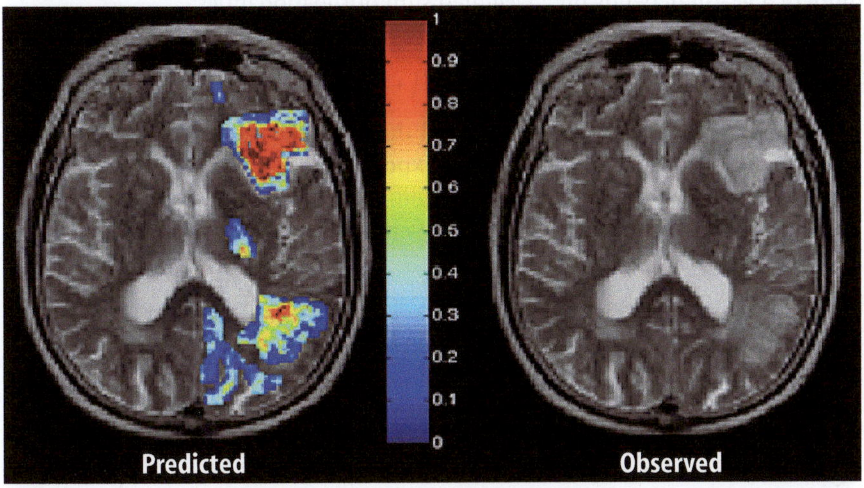

Fig. 6. Using acute perfusion and diffusion images from a patient presenting with symptoms of acute stroke, the risk of tissue infarction was calculated in each voxel using a logistic regression model. The figure shows the estimated risk map (*left*) and the 3-month follow up T2 image (*right*). The model correctly predicts infarction of the anterior part of the territory supplied by the left middle cerebral artery and the adjacent anterior and posterior watershed areas

predicted disease progression under different therapeutic strategies and for assessing treatment efficacy.

In other situations, the outcome may be continuous (e.g., MTR change). In the previous sections, we described how to regress observations sampled at one time point on observations from another time point. Considering the first measurement as a predictor for the second measurement, this is a simple example of a situation where the outcome – the measurement at the second time point – is continuous. If the exact value of an outcome is not important, but only the position relative to some threshold, the regression model can be converted into a classification problem, which may be modeled using logistic regression. This can be advantageous because linearity and distributional assumptions become unnecessary.

The inference techniques described in the following sections can be used regardless of the scale of the outcome, but we describe the procedures in relation to classification problems, since many concepts are most easily explained in this setting. The only exception is the section on Predictive Ability, which is primarily relevant for classification problems. (For a thorough treatment of regression modeling for both binary and continuous outcomes, see [5]).

Predicting Outcome

Considering a single voxel, we assume the values of the k modalities are $x_1, \ldots, x_k$. Then we wish to model, e.g., the probability of a T2 lesion in this voxel within a fixed number of years. Using the standard logistic regression model, the probability of a lesion depends on a weighted sum of the modalities:

$$P(T_2 \text{lesion} \mid x_1, \ldots, x_k) = \frac{1}{1 + \exp(-\sum_{i=1}^{k} \alpha_i x_i)} \quad .$$

Suppose that x_1 represents MTR, and the fitted coefficient α_1 is negative. If MTR decreases, this means that the exponential term in the denominator becomes small and therefore the probability of a lesion increases. In particular, the odds of a lesion developing are represented by:

$$\text{odds}\ \{T_2 \text{ lesion}\} = \frac{P(T_2 \text{ lesion} \mid x_1, \ldots, x_k)}{P(No\ T_2 \text{ lesion} \mid x_1, \ldots, x_k)} \qquad (3)$$

Logistic regression models are used exactly as the typical linear regression model except that the latter is used to model a continuous outcome. In fact the logistic model is also a linear model in the sense that if we take the logarithm on both sides in Eqn (3):

$$\log \text{odds}\ \{T_2 \text{ lesion}\} = \sum_{i=1}^{k} \alpha_i x_i = \alpha_1 x_1 + \alpha_2 x_2 + \ldots + \alpha_k x_k ,$$

it can be seen that the logistic regression model corresponds to a linear regression with the response transformed to the log-odds scale.

Identifying Relevant Modalities

Often, in practical situations, many predictor variables are entered into the initial model. However, there may be reasons to attempt to reduce this model. For example, some of the involved imaging procedures may be time-consuming or difficult to perform, and therefore it is desirable to test whether such modalities actually contribute to predicting the outcome. It may also be that some of the images reflect the same physiological property. Some of the variables are therefore redundant and should be removed to avoid collinearity, which inflates the standard errors of the estimates of the coefficients. Moreover, less complex models are easier to interpret. On the other hand, reducing the model may result in an increase in bias. Therefore, parsimony and accuracy should be balanced when simplifying a model.

Significance tests for the hypotheses that certain coefficients are equal to zero (i.e., that a certain modality does not contribute additional information to the likelihood of an outcome) can be performed exactly as in linear regression. These tests are based on the likelihood ratio, score, or Wald statistics. When testing these hypotheses using statistical packages, the likelihood ratio test should be first choice. When the true value of a parameter is numerically much larger than zero, the Wald statistic tends to zero and hence a significant factor will be considered to be not significant [24].

When multiple imaging modalities are entered into a model, it is typical to commence stepwise variable selection to reduce the model one variable at the time. Although widely used, stepwise variable selection has several serious drawbacks including: (a) the standard errors of the coefficient estimates are biased low, (b) adjustment for multiple comparisons is necessary, but proper correction is difficult, and (c) the estimated coefficients are biased high. Therefore, the final model should be interpreted with care.

Instead of randomly evaluating all possible models, insight into subject matter should determine the order in which variables should be tested. When the model cannot be reduced any further after a series of tests, a single, overall test from the full model to the reduced model should be performed to confirm the findings. One additional technique to protect against selection of nonsignificant variables is to begin the analysis with a test from the full model to the model with a single global mean (no regression). If this test is not significant, none of the variables in the full model should be considered significant.

There are a number of automatic procedures performing forward or backward stepwise variable selection based on either P values, Akaike's information criteria, Bayes information criteria, or Mallow's C_p. A very efficient method based on Wald statistics has also been suggested [25], and is implemented in S-PLUS. To adjust for the bias toward high values of the coefficient estimates, a penalized estimation technique (shrinkage methods) can be used to estimate in the final model. Ridge regression [26] and the Lasso [27] are implemented in S-PLUS. Another possibility is to employ parsimony constraints directly in the model-building process. One such procedure is multivariate adaptive regression splines [28], which use a generalized cross-validation criteria to penalize complex models.

Dependence between Predictors

When several modalities are used to predict an outcome, there is a risk that some of the predictors are not independent. Some modalities may even be algebraically related. Generally this is problematic because it may inflate the variance of coefficient estimates, decrease the power of associated tests [29], and lead to incorrect conclusions about which variables are important. While pairwise dependence – and algebraic dependence – can be detected simply by plotting each predictor against the others, non-trivial dependencies involving more than two predictors are more difficult to identify. A formal way to quantify the degree of dependence is to essentially divide the actual variance of a coefficient by the variance it would have if the predictor were independent of all other variables [30]. This measures how much the observed variance is inflated by collinearity, and the ratio is called the variance inflation factor (VIF). In practice, the VIF of a predictor is calculated as $1/(1 - R_a^2)$, where R_a^2 is the squared (adjusted) multiple correlation coefficient for the dependence between this and all other predictors. If the model is fitted using maximum likelihood, VIFs are calculated based on Fisher's information matrix. Both SAS and S-PLUS offer functions to calculate VIFs.

As a rule of thumb, if the VIF of a predictor exceeds a value of 10, it is likely involved in multicollinearity. Once a group of collinear predictors is identified, attempting to rank them according to importance is not recommended. Instead, their joint contribution should be assessed by testing if they can, simultaneously, be excluded from the model. If this test is significant, the predictors should be combined into a single score (such as the average), which will replace the individual predictors. Note that both SAS and S-PLUS have functions for identifying hierarchical groups of predictors.

Relative Importance of Individual Modalities

This section deals with assigning importance to image modalities in the disease progression models – and thereby indirectly assigning importance to the underlying biophysical changes in the disease process. The hypothesis tests described above are concerned with removing insignificant predictors from the model. When a model cannot be further reduced, a natural question is whether it is possible to rank the modalities according to their relative importance. Although conceptually clear, it is strikingly difficult to reach consensus on a formal definition of this issue, and some authors dispute the meaning and appropriateness of assessing relative strength of a predictor altogether (see [31] for a list of references). One study of the literature [32] found substantial diversity in applied measures, but concluded that many were too simplistic, considering only isolated effects of single variables, and, critically, about 20% of the papers used *P* values as measures of relative importance. The use of P-values as a measure of relative importance is strongly discouraged. A P-value pertains to the sample as well as the population, whereas relative importance is solely a population property.

Traditional measures of relative importance include standardized regression coefficients and marginal correlation coefficients. Standardized regression coefficients are obtained simply by multiplying the original coefficient by the ratio between the standard deviation of the respective predictor and the standard deviation of the response. Standardized coefficients are, in contrast to the original coefficients, comparable because they represent the change in standard deviation units of the response variable for a change of one standard deviation in the predictor variable. Marginal correlation coefficients are used because of the interpretation that they quantify the proportion of variation in the response variable explained by the individual predictors.

In the particular case where there is a well-defined ordering of the predictors, for instance if they represent different time points, defining relative importance is less problematic. The relative importance of the first predictor is the marginal correlation coefficient, but the relative importance of the second predictor is the proportion it accounts for of the *remaining* variation, and so forth with the rest of the predictors. Recently, several authors [31, 33, 34] have suggested effectively using the average relative importance for a predictor over all orderings as a measure of the overall strength of a predictor. Another recent approach to quantifying the relative contribution of a predictor is based on proportional marginal decomposition (see http://www.prismanalytics.com/Papers.htm). Essentially, this procedure also considers all possible orderings, but additionally it evaluates the probability that a given ordering reflects the relative importance of predictors. Both of these recent approaches are implemented in S-PLUS via the RELIMPO package.

The Strength of a Model: Measuring Predictive Ability

Measures of Predictive Ability

When a prognostic model has been developed, it is of key importance to quantify its ability to correctly estimate the probability of an outcome based on future data. There are a host of indices of predictive ability, many of which are equivalent but appear under different names. A general advice is to consider measures which have a clear interpretation and do not require fixing a probability threshold used to allocate observations into the two groups based on whether the probability of the outcome exceeds the cutoff point.

A widely used index is the area under the receiver operating characteristic curve (ROC). This is equivalent to the probability of concordance (the *c* index) between predicted probability and response, and it is linearly related to Somers' rank correlation. It is effectively a measure of the capability to discriminate between outcomes. If we are modeling the probability of a T2 lesion based on imaging modalities, then the area under the ROC curve, denoted as AUC, is the probability that a voxel, in which a lesion developed, has a higher probability score than one which in which a lesion did not develop. Loosely phrased, the AUC is the probability of correct ranking. As an additional benefit, the ROC curve is based on the two well-known quantities, *sensitivity* and *specificity*. For a particular threshold, the sensitivity is the proportion of voxels developing lesions which were predicted by the model, and the specificity is proportion of voxels which did not develop a lesion which were correctly predicted by the model. The ROC curve is created by plotting sensitivity against (1 – specificity) for all values of the threshold. A value of the AUC close to 0.5 indicates random predictions, and a value close to 1 indicates perfect prediction.

Other measures of predictive value include: (a) logarithmic scoring (Shannon's entropy) and quadratic scoring (Brier's score), which are slightly harder to interpret than AUC, (b) the probability of correct classification (accuracy), which depends on a specific threshold, and (c) the generalized R_N^2 [35, 36], which is a standardization of the Maddala index [37].

Estimating Predictive Ability

As pointed out in the previous section, there is a risk of over-fitting a model. Therefore goodness-of-fit indices, measuring the prediction error on the same data for which the parameters were optimized, will be biased low relative to how well the model will fit future observations. Consequently, the prediction error which is based on the original data is termed the apparent prediction error. Ideally, predictive performance should be estimated by applying the model to another data set. Clearly such a data set will rarely be available, but a good alternative is to divide the data at hand into a training set and a test set. The model is then estimated – or trained – using the training arm of the original data, and a measure of reliability is obtained by using the model to predict the outcomes in the test set. However, this estimate will depend on the actual split of the data, and therefore the procedure should be repeated for other partitions of the original data. This is exactly the rationale behind cross-validation. Here the data is divided into K subsets, and the procedure begins by setting aside the first subset for testing and the model is trained on the remaining K–1 groups. After calculating model performance on the test set, another one of the K subsets is appointed test set and the procedure is repeated K times. The final estimate of model performance is the average of the K prediction error estimates. A special case of cross-validation obtains when the number of groups equals the number of observations. This is known as leave-one-out cross-validation, where each observation is sequentially left out. However, larger groups give more accurate results [38].

Estimates of prediction error obtained by cross-validation depend on the particular grouping employed, and critically, this method does not validate the full model but only models based on subsets of the original data. It has also been demonstrated that cross-validation, although roughly unbiased, can exhibit large variability [39: p. 255]. The bootstrap [39] is a very powerful generalization of cross-validation. The principal idea is to randomly sample (with replacement) observations from the original data to obtain a training set instead of selecting a fixed group as in cross-validation. Although this represents some improvement over traditional cross-validation, a superior approach is to use the bootstrap to correct the apparent prediction error. This is done by iteratively training the model on the bootstrap sample and then, counter-intuitively, estimating prediction error by applying this model to the original data including the bootstrap sample. This estimate will be biased low, but if we subtract the prediction error within the bootstrap sample, we get an estimate of the "optimism", which is the amount by which the apparent prediction error underestimates the actual prediction error. The apparent prediction error, estimated using a full model, can now be corrected by adding the average "optimism" estimated using the bootstrap. This estimate is less biased downward than simple bootstrapping and shows less variability than cross-validation (see [39] p. 255], which also describes an additional improvement).

The S-PLUS package, and to some extent SAS, offers routines for obtaining cross-validation and bootstrap estimates of prediction error.

Longitudinal Studies

Many studies track evolutions in image modalities over time. One of the virtues of the linear regression model introduced above is that it can easily be adapted to repeated-measure studies. The only necessary alteration is the specification of the correlation structure implied by the temporal dependence. Suppose we observe the values $y_1, y_2, \ldots, y_m$ in a voxel at time points $t_1, t_2, \ldots, t_m$. In the typical regression model, we assume that the errors $\varepsilon_1, \varepsilon_2, \ldots, \varepsilon_m$ are independent. However, when observations are acquired serially, it is natural to assume that observations are correlated such that $cor(y_i, y_j) = \rho_{ij}$. This general correlation structure can be simplified by assuming that the correlation only depends on the temporal distance $|t_i - t_j|$, in which case: $cor(y_i, y_j) = \rho(|t_i - t_j|)$ Observations are often assumed to correlate more with "neighboring" observations than observations made at much later or earlier time points. The autoregressive model assumes that the error at time t_i depends linearly on the p previous errors:

$$\varepsilon_i = a_1\varepsilon_{i-1} + \ldots + a_p\varepsilon_{i-p} + \phi_i.$$

Here ϕ_i is a zero-mean random variable with constant variance. In the case where the error in an observation only depends on the previous error, we have:

$$cor(y_i, y_j) = a^{|t_i - t_j|}, \quad a \geq 0.$$

The autoregressive model was introduced for integer time points, but it can be extended to continuous time measurements. (See [40, 41] for more details on longitudinal data analysis.)

Measuring Therapeutic Efficacy using Predictive Models

In this section, we discuss the application of predictive models in measuring therapeutic efficacy.

We have discussed methods for measuring treatment efficacy for continuous outcomes in the section on Hierarchical Models. In this section, we extend this to situations where the probability of an outcome is modeled. Predictive models are particularly useful for measuring the efficiency of a treatment in the sense that effects can readily be interpreted as increases or decreases in the probability of the outcome for identical values of input parameters. (See [23] for an example of the use of predictive models measuring the efficacy of thrombolytic therapy in stroke.)

A simple procedure for quantifying treatment effects is to include a dummy variable in the logistic regression model, which identifies observations pertaining to the treated group:

$$P(outcome \mid z, x_1, \ldots, x_k) = \frac{1}{1+\exp\left\{-\left(\alpha z + \sum_{i=1}^{k} \beta_i x_i\right)\right\}}$$

Here z has the value 1 for observations from treated patients and zero otherwise. Then the parameter α quantifies the effect of the treatment. If α is large and positive, then the treatment increases the probability of the outcome for identical values of the predictors. This means that if a patient from the treated group and a patient from the untreated group have identical values of, for example, MTR and diffusion metrics, then the patient from the treated group has an increased probability that the outcome of interest – e.g., reversal of the evolution of a T2 lesion – will occur.

One drawback of this method is that it models treatment efficacy as a constant, global change in probability. If the changes in sensitivity to values of the imaging modalities induced by a drug are more complex, the value of α may be small, implying that the treatment was inefficient. An alternative strategy, which applies to general prognostic algorithms, is to build separate models for the treated and the untreated patients. This way, we can estimate the probability of the outcome $P_{treated}(outcome \mid x_1, \ldots, x_k)$ for treated patients, which should be compared to the estimated probability $P_{untreated}(outcome \mid x_1, \ldots, x_k)$ for untreated patients. If the treatment is speculated to increase the probability of the outcome, we should assess whether:

$$P_{untreated}(outcome \mid x_1, \ldots, x_k) < P_{treated}(outcome \mid x_1, \ldots, x_k)$$

for all values of $x_1, \ldots, x_k$. This can easily be checked graphically by plotting the two probabilities for all observed values of $x_1, \ldots, x_k$. If the relation is approximately linear, the treatment effect may be quantified by the intercept and slope parameters of a fitted regression line.

An important quality of this approach is that if a model with high predictive accuracy can be built for patients receiving no therapy, then this model can be applied in new patients, before treatment is initiated, to determine disease progression if no treatment is given. Hence, predictive models can be used in the planning of therapeutic strategies, since the effects of different treatment scenarios can be estimated. Therapeutic efficacy can be concurrently assessed by measuring the discrepancy between actual outcomes after treatment and outcomes predicted with the model trained on patients receiving no treatment. In cases where the outcome is modeled for each voxel, this method can also be employed in order to evaluate regional effects of possible treatments. By applying the model for the treatment group as well as the model for the untreated group to data from a new patient, the difference in probability of the outcome of interest can be calculated for each voxel. These differences can then be superimposed on an anatomical image to give a color-coded map of regional treatment effects.

Discussion

Because of the wealth of multimodal data acquired in biomedical research and drug trials related to MS, data-mining techniques are crucial in order to fully exploit the power of these data in detecting disease- or treatment-related changes, or generating novel hypotheses. Through the development of sensitive MR-based biomarkers to detect diffuse as well as focal changes, MS research is ideally positioned to take advantage of recent progress in the development of powerful statistical modeling tools. Figure 7 summarizes techniques for common data types in multimodal imaging; most of which we have sought to address in this chapter.

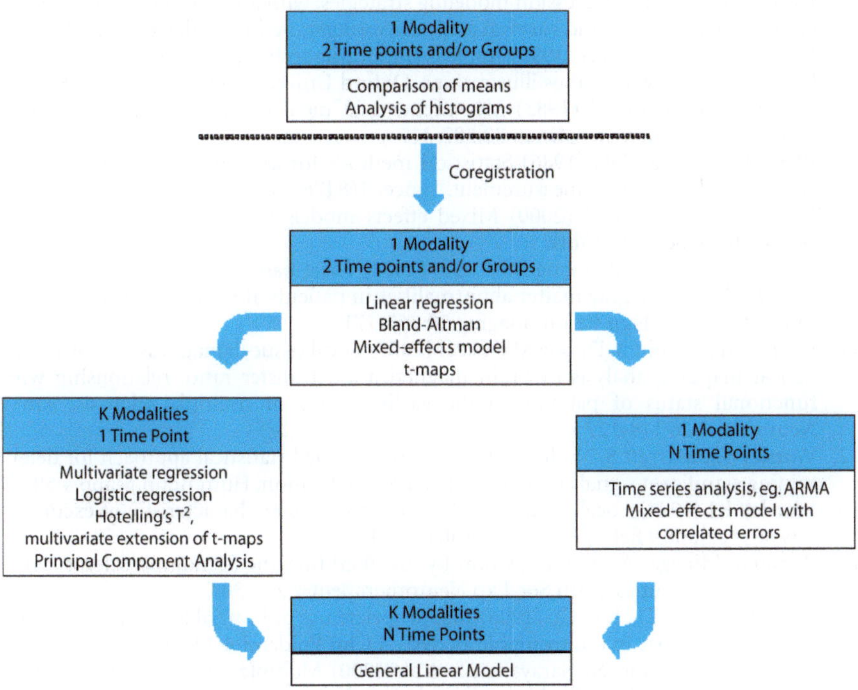

Fig. 7. Suggested analysis strategies for some common imaging studies

Note on Software

The examples presented in this review were analyzed using R (http://www.r-project.org/), which is a freely available implementation of the S language, which also underlies S-PLUS. Hence, most code developed in S-PLUS will run in R and vice versa.

Acknowledgements. The authors wish to thank Professor Jens Ledet Jensen from the Department of Theoretical Statistics, Aarhus University, and Kristjana Fonsdottir from the center for Functionally Integrative Neuroscience, Aarhus University Hospital, for helpful comments and suggestions.

References

1. Collins DL, Neelin P, Peters TM, Evans AC (1994) Automatic 3D intersubject registration of MR volumetric data in standardized Talairach space. J Comput Assist Tomogr 18:192-205
2. Fischl B, Sereno MI, Tootell RBH, Dale AM (1999) High-resolution intersubject averaging and a coordinate system for the cortical surface. Hum Brain Mapp 8:272-284
3. van Buchem MA, McGowan JC, Kolson DL et al (1996) Quantitative volumetric magnetization transfer analysis in multiple sclerosis: estimation of macroscopic and microscopic disease burden. Magn Reson Med 36:632-636
4. Mardia KV, Kent JT, Bibby JM (1979) Multivariate analysis. Academic Press, London
5. Harrell FE Jr (2001) Regression modeling strategies: with applications to linear models, logistic regression and survival analysis. Springer, Berlin Heidelberg New York
6. Bowman AW, Azzalini A (1997) Applied smoothing techniques for data analysis: the kernel approach with S-Plus illustrations. Oxford University Press, Oxford, UK
7. Altman DG, Bland JM (1983) Measurement in medicine: the analysis of method comparison studies. Statistician 32:307-317
8. Bland JM, Altman DG (1986) Statistical methods for assessing agreement between two methods of clinical measurement. Lancet 1(8476):307-310
9. Pinheiro JC, Bates DM (2000) Mixed-effects models in S and S-PLUS. Springer, Berlin Heidelberg New York
10. Audoin B, Ranjeva JP, Duong MVA et al (2004) Voxel-based analysis of MTR images: a method to locate gray matter abnormalities in patients at the earliest stage of multiple sclerosis. J Magn Reson Imaging 20:765-771
11. Ranjeva JP, Audoin B, Duong MVA et al (2005) Local tissue damage assessed with statistical mapping analysis of brain magnetization transfer ratio: relationship with functional status of patients in the earliest stage of multiple sclerosis. Am J Neuroradiol 26:119-127
12. Worsley KJ, Marrett S, Neelin P et al (1996) A unified statistical approach for determining significant signals in images of cerebral activation. Hum Brain Mapp 4:58-73
13. Worsley KJ (1994) Local maxima and the expected Euler characteristic of excursion sets of χ^2, F and t fields. Adv Appl Probab 26:13-42
14. Bakshi R, Minagar A, Jaisani Z, Wolinsky JS (2005) Imaging of multiple sclerosis: role in neurotherapeutics. J Am Soc Exp Neurotherapeut 2:277-303
15. Sun GW, Shook TL, Kay GL (1996) Inappropriate use of bivariable analysis to screen risk factors for use in multivariable analysis. J Clin Epidemiol 49:907-916
16. Pike GB, De Stefano N, Narayanan S et al (2000) Multiple sclerosis: magnetization transfer MR imaging of white matter before lesion appearance on T2-weighted images. Radiology 215:824-830

17. Fazekas F, Ropele S, Enzinger C et al (2002) Quantitative magnetization transfer imaging of pre-lesional white-matter changes in multiple sclerosis. Mult Scler 8:479-484
18. Laule C, Vavasour IM, Whittall KP et al (2003) Evolution of focal and diffuse magnetisation transfer abnormalities in multiple sclerosis. J Neurol 250:924-931
19. Rocca MA, Cercignani M, Iannucci G et al (2000) Weekly diffusion-weighted imaging of normal-appearing white matter in MS. Neurology 55:882-884
20. Santos AC, Narayanan S, De Stefano N et al (2002) Magnetization transfer can predict clinical evolution in patients with multiple sclerosis. J Neurol 249:662-668
21. Agosta F, Rovaris M, Pagani E, Sormani MP et al (2006) Magnetization transfer MRI metrics predict the accumulation of disability 8 years later in patients with multiple sclerosis. Brain 129:2620-2627
22. Wu O, Koroshetz WJ, Østergaard L et al (2001) Predicting tissue outcome in acute human cerebral ischemia using combined diffusion- and perfusion-weighted MR imaging. Stroke 32:933-942
23. Wu O, Christensen S, Hjort N et al (2006) Characterizing physiological heterogeneity of infarction risk in acute human ischaemic stroke using MRI. Brain 129:2384-2393
24. Hauck WW, Donner A (1977) Wald's test as applied to hypothesis testing in logit analysis. J Am Stat Assoc 72:851-853
25. Lawless JF, Singhal K (1978) Efficient screening of nonnormal regression models. Biometrics 34:318-327
26. Lecessie S, Vanhouwelingen JC (1992) Ridge estimators in logistic-regression. J R Stat Soc Ser C 41:191-201
27. Tibshirani R (1996) Regression shrinkage and selection via the Lasso. J R Stat Soc Ser B 58:267-288
28. Jerome F (1991) Multivariate adaptive regression splines. Ann Stat 19:1-141
29. Glantz SA, Slinker BK (1990) Primer of applied regression and analysis of variance. McGraw-Hill, New York
30. Gross J (2003) Linear regression. Springer, Berlin Heidelberg New York
31. Chevan A, Sutherland M (1991) Hierarchical partitioning. Am Stat 45:90-96
32. Kruskal W, Majors R (1989) Concepts of relative importance in recent scientific literature. Am Stat 43:2-6
33. Kruskal W (1987) Relative importance by averaging over orderings. Am Stat 41:6-10
34. Soofi ES, Retzer JJ, Yasai-Ardekani M (2000) A framework for measuring the importance of variables with applications to management research and decision models. Decision Sci 31:595-625
35. Nagelkerke NJD (1991) A note on a general definition of the coefficient of determination. Biometrika 78:691-692
36. Cragg JG, Uhler RS (1970) Demand for automobiles. Can J Econ 3:386-406
37. Maddala GS (1983) Limited-dependent and qualitative variables in econometrics. Cambridge University Press, Cambridge, UK
38. Efron B (1983) Estimating the error rate of a prediction rule: improvement on cross-validation. J Am Stat Assoc 78:316-331
39. Efron B, Tibshirani RJ (1993) An introduction to the bootstrap. Chapman and Hall, London, pp 247-249
40. Box GEP, Jenkins GM, Reinsel GC (1994) Time series analysis: forecasting and control. 3rd edn. Holden-Day, San Francisco, CA
41. Jones RH (1993) Longitudinal data with serial correlation: a state-space approach. Chapman and Hall, London

LESSONS FROM OTHER NEURODEGENERATIVE DISEASES

Chapter 13

Alzheimer's Disease

G.B. Frisoni

Introduction

Clinical trials of Alzheimer's disease (AD) traditionally use rating scales, such as neuropsychological tests, and disability scales as outcome measures. However, their intrinsic measurement variability, the slow disease progression, and the low efficacy of the drugs developed so far have led to trial designs with hundreds of subjects per treatment arm. The development of imaging markers with proven sensitivity to disease progression has recently paved the way for their use as outcome measures in clinical trials. The use of imaging measures has the double advantage of decreasing the number of subjects per treatment arm whilst also providing a direct measure of the degree of disease modification induced by the "active" molecules. A number of magnetic resonance (MR)-based markers have been developed for clinical trials of AD, all of which have their own strengths and weaknesses. Here, the most often used techniques, which could easily be exported to the study of neurodegeneration in clinical trials of multiple sclerosis, are reviewed.

Region-of-Interest-Based Volumetric Measures

After acquisition, the digital images need to be reconstructed on coronal, 1-2 mm-thick slices. The hippocampus is then manually traced on all the contiguous slices where it can be seen (Fig. 1).

In expert hands, the reliability is high, intra-class correlation coefficients for hippocampal measurements being 0.95 for intra-rater and 0.90 for inter-rater variability [1]. Sensitivity and specificity values of hippocampal volumes in a relatively large series of 55 AD patients and 42 controls were 94% and 90%, respectively [2]. Small hippocampal volume was found to be predictive of subsequent conversion to AD in 80 patients with amnestic mild cognitive impairment (MCI) independently of neuropsychological tests, apolipoprotein E genotype, and cerebrovascular comorbidity [3]. Of the 13 MCI patients with hippocampal volume 2.5 SDs below the age-specific mean, only 6 (46%) converted to AD within the following 6 years; while of the 54 with hippocampal volume between –2.5 SDs and the age-specific mean, converters were 19 (35%); and of the 13 with hippocampal volume above the mean, only 2 (15%) converted [3].

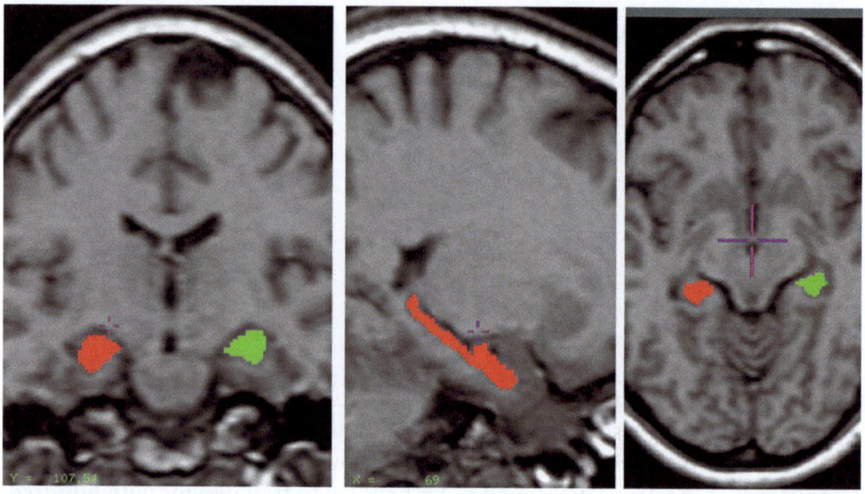

Fig. 1. Regional atrophy assessment: hippocampal volumetry. An example of manual tracing of the right hippocampus with simultaneous view of the traced region-of-interest in the coronal, sagittal, and axial planes (*red*). The left hippocampus (*green*) is seen on the coronal and axial images

In a prospective study of 28 patients with AD, 43 with MCI, and 58 normal controls, the annualized percentage change in hippocampal volume decreased progressively, according to expectation, from baseline cognitive status (AD>MCI>control) [4]. Moreover, within the control and MCI groups, the rate of change in hippocampal volume correlated with the change in cognitive status over time (control-stable=1.73% vs control-decliner=2.81%; MCI-stable=2.55% vs MCI-decliner=3.69%). This study suggests that serial measurements of the hippocampal volume allow us to detect stable versus declining members of a group, and may therefore be a useful tool for monitoring the efficacy of therapeutic trials.

Computational Neuroanatomy

Recently, advances in neuroscience and neuroimaging have led to an increasing recognition that certain neuroanatomical structures may be affected preferentially by specific diseases. Neurodegenerative brain diseases mark the brain with a morphological "signature"; detecting this may be useful for enhancing diagnosis, particularly in diseases where there is a lack of other diagnostic tools. Moreover, structural changes provide markers which enable us to track the biological progression of the disease.

Recent developments in computer science may contribute to the detection of early, sensitive, and specific disease signatures. The new approaches are automated, thus avoiding error-prone and labor-intensive manual measurements. Such algorithms can also offer unprecedented precision, as some can detect brain volume differences of 0.5% between images from the same subject [5]. The research work needed in order to develop such algorithms is referred to as computational neuroanatomy [6].

The individual algorithms can be categorized into two broad classes: algorithms devised to detect group differences at one point in time, and algorithms devised to detect prospective changes over time. The first category may be useful for defining disease-specific signatures. The second can be applied to one or more individuals to track disease progression, either natural or modified by treatment. While most tools have been developed to compare groups, some are being adapted to analyze individual cases; an issue of the greatest interest for the practicing physician.

Computational anatomy algorithms generally involve some or all of the following steps: (a) brain extraction (brain is separated from non-brain voxels), (b) tissue segmentation (voxels representing gray matter and white matter and cerebrospinal fluid are separated, based on intensity values), (c) spatial normalization (also called registration; the voxels of interest are matched to a template or an earlier scan from the same individual), and (d) statistical comparison of different subject groups or points in time. The pivotal step for all these methods is registration. Here, cross-sectional methods match images of interest to a reference stereotactic template (a typical brain or a typical hippocampus, etc.) or vice versa, while prospective methods match sequential images from the same patient taken at different times.

Registration strategies differ in scope (i.e., analysis of the whole brain or preselected regions-of-interest) and mathematical approach (accounting for global or local variability in the brain's size and shape). Some cross-sectional methods that account for global variability are completely automated (such as voxel-based morphometry based on statistical parametric mapping, as developed by Ashburner and Friston – A&F) [7], while those that account for local variability often require manually positioned landmarks to precisely match the image to the template (such as cortical pattern matching) [8]. Longitudinal methods use the complexity of each individual's brain structure to align accurately an individual's serial images (such as the brain boundary shift integral – BBSI – algorithm) [9].

To perform well, all methods need high spatial resolution and clear differentiation between tissue types. Usually, 3D, high-resolution, T1-weighted MR images (spoiled gradient-recalled – SPGR – or magnetization-prepared rapid acquisition gradient-recalled echo – MP-RAGE) acquired with conventional 1.5T MR scanners and 1 mm3 voxels (ideally isotropic) provide sufficient detail and contrast (Fig. 2).

Voxel-Based Morphometry (VBM) (Ashburner and Friston's Method)

Here, registration is made to a global template, which allows for the removal of the global, but not the local, shape variability. Registered gray matter images are smoothed with an 8-12 mm filter, leading to normally distributed data and allowing the use of statistical parametric tools. A *t* test is then performed on a

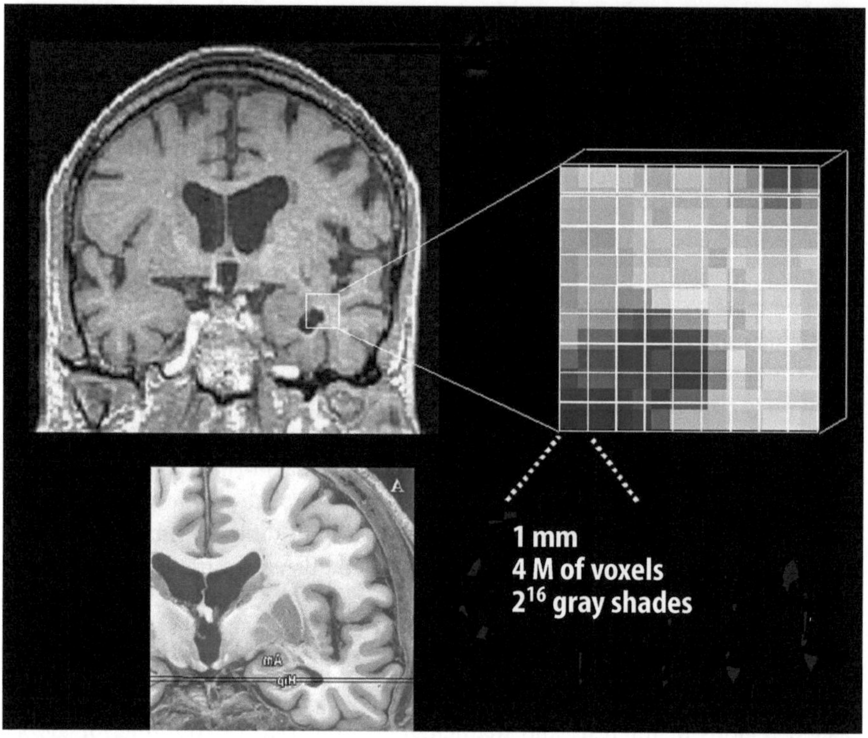

Fig. 2. High-resolution, T1-weighted, 3-dimensional (MP-RAGE or SPGR) techniques are used where whole-skull images are made of voxels of about 1 mm^3 within which tissue density is homogeneous (isotropic). The whole volume is made up of about 4M voxels. Each voxel can generally have 2^{16} shades of gray

voxel-wise basis between groups of subjects or within a group of subjects scanned at baseline and follow-up (Fig. 3).

The statistical approach of the A&F method Statistical Parametric Map (SPM) is based on the general linear model, and identifies regions of tissue with increased or decreased density or concentration that are significantly related to the effects under study. Ideally, the threshold for significance should be set at $p<0.05$ corrected for multiple comparisons, but when there is an a priori hypothesis of the expected effect, a more liberal threshold of $p<0.001$ uncorrected can be used. However, like every statistical test, the larger the effect size and group size, the higher the sensitivity of the method for identifying differences. The A&F method has been implemented in software running under Microsoft Windows or UNIX that can be downloaded freely (http://www.fil.ion.ucl.ac.uk/spm/).

In a recent prospective study, Chetelat et al. [10] studied 18 patients with MCI for 18 months. At the end of the follow-up period, 7 had converted to probable AD, and 11 remained MCI. All patients underwent two high-resolution MR examinations at baseline and at follow-up, and MR images were processed and

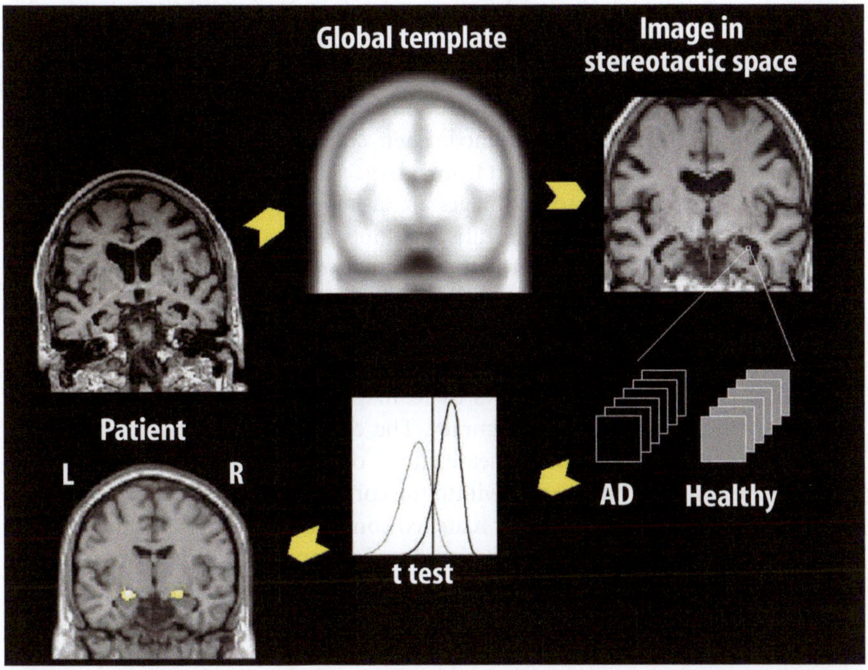

Fig. 3. Voxel-based morphometry: each original image is registered to a global template, which allows the global shape variability to be removed. Registered gray matter images are smoothed with an 8-12 mm filter (not shown), leading to normally distributed data and allowing the use of statistical parametric tools. A *t* test is then performed on a voxel-wise basis between groups of subjects or within a group of subjects scanned at baseline and follow-up and a voxel-wise significance map is obtained

analyzed with SPM2 following a longitudinal VBM protocol based on individual high-dimensional warping of the follow-up onto the baseline image. Gray matter loss was estimated from the Jacobian determinants of the applied deformations. Gray matter loss in both converters and non-converters was located in the lateral and medial temporal areas, bilaterally, the orbitofrontal and inferior parietal areas, and the left thalamus. Converters showed greater gray matter loss than non-converters in the temporal neocortex, medial temporal lobe structures, posterior cingulate, and precuneus, bilaterally. Percentage annual losses ranged between 0% and 4.5% for converters, and between 0% and 4% for non-converters, with the temporal lobe and the lateral temporal cortex being involved at the highest rates.

Cortical Pattern Matching

This is a sensitive approach that measures the topological variability of the cortex [11]. The approach consists of cortical flattening and sulcal matching that aim to obtain an average cortical model for a group of subjects. All MR scans are first aligned to a standardized 3D coordinate space, then from each individual's MR scan, a 3D cortical surface model is extracted, consisting of a network of discrete triangular tiles, and some sulcal/gyral landmarks are identified on the cortical model (Fig. 4), which is then flattened. Then, sulcal features are co-registered to a template of sulcal curves, derived from a large group of normal subjects, with a warping technique. The warped images are averaged and measures of gray matter density can be analyzed with statistical tools similar to those used by A&F's VBM. This technique can be performed on high-end desktop machines such as the Macintosh G4, as well as Silicon Graphics Interface or Sun workstations running UNIX. These algorithms are often used in client-server mode, connecting to a supercomputer for very large-scale analyses.

Cortical pattern matching allows us to map changes of cortical gray matter density or thickness with great accuracy. The cortical pattern matching analysis has been used to better localize disease effects on cortical anatomy over time. To visualize the transit of the disease within the cortex, Thompson et al. [12] studied 12 patients with AD and 14 elderly matched controls scanned longitudinally (two scans). In patients with AD, a highly significant gray matter loss was observed in a broad anatomical region encompassing bilateral temporal and parietal cortices. The most significant changes occurred in temporal and parietal regions, where tissue loss exceeded 15%. Primary sensory and motor cortices were comparatively spared in the disease (with a 0%-5% deficit, on average, in the central and postcentral gyri). After 1.5 years, the frontal cortices, initially only mildly affected, were found to be severely affected (tissue loss exceeded 15%), whereas the sensory and

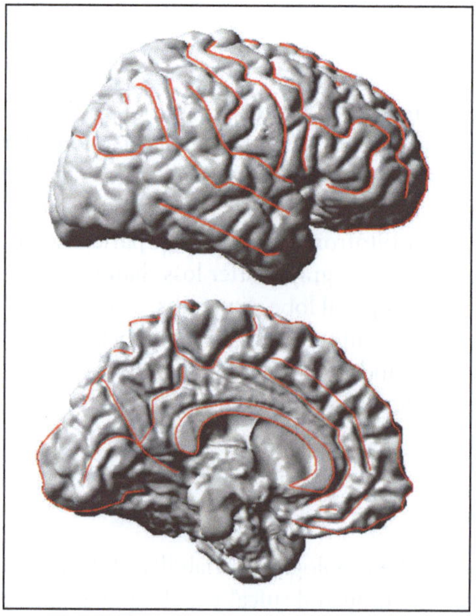

Fig. 4. Cortical pattern matching: sulci that need to be traced manually on the medial and lateral surfaces in order to carry out cortical registration that will remove global as well as local variability

motor territories were still relatively spared. This study provides the first quantitative, dynamic visualization of cortical atrophy rates in dementia.

Vidal et al. [13] studied 12 schizophrenic subjects and 12 healthy volunteers. All subjects were studied prospectively during a 5-year period. To detect earlier loss, they compared gray matter profiles across all 24 subjects at their first and at their last scan 4.5 years later. Medial frontal cortex was affected early in the disease course from the anterior frontal regions to the posterior limit of the precentral gyrus in both right and left hemispheres. This damage increased at follow-up in each hemisphere. The medial parietal cortex was also significantly affected early and late in the course of the disease. Importantly, the cingulate was relatively well preserved at the onset of the disease, but was significantly affected at follow-up, in the left hemisphere. Frontal deficits, which are characteristic of adult and childhood schizophrenia, were severely progressive, but they were already present in the early phase of the disease.

Brain Boundary Shift Integral (BBSI)

Serial scans of the same subjects have the advantage that the wide inter-individual variability of brain morphology is not an issue, and comparing pre-post images of the same subject(s) carries much less error than comparing a case to a control. Information on prospective global changes can be obtained by rigidly matching serial scans and by subtracting the superimposed images. The difference reflects the volume of brain tissue lost or gained (e.g., brain boundary shift integral, BBSI [9]) (Fig. 5). The rate of atrophy in a group of 18 AD patients was significantly greater than in 31 controls (2.78% vs 0.24% per year with no overlap between the groups) [5]. Moreover, the rate of global cerebral volume loss was strongly correlated with rate of cognitive impairment measured with the mini-mental state examination (MMSE) in 29 AD patients [14].

Information on local changes can be obtained by using nonlinear registration, which permits compression or expansion of each voxel to obtain a precise

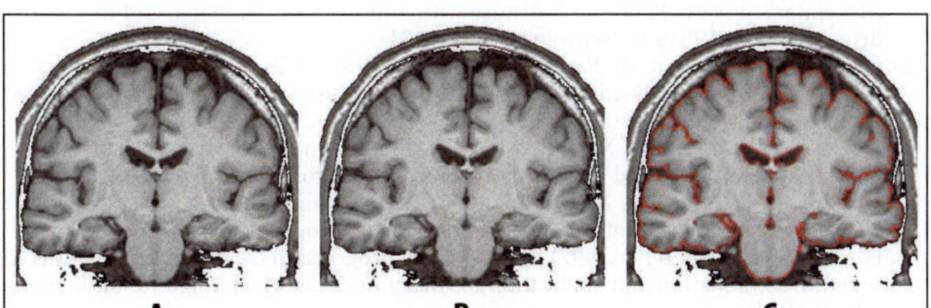

Fig. 5. Brain boundary shift integral: brain MR scan of a patient with AD at baseline (**A**), at follow-up (**B**) (no gross differences can be perceived), and the subtraction image (**C**), where voxels of atrophy between baseline and follow-up are shown in *red*

registration (voxel compression method). The resulting deformation fields provide a map (voxel compression map – VCM) of the amount of compression or expansion applied at each voxel. This reflects the amount of brain tissue or cerebrospinal fluid lost or gained over time in the interval between scans. Typical patterns of change in different conditions (e.g., normal aging vs AD) can be tapped by registering and averaging individual voxel compression maps and comparing the resulting averages. This method was capable of detecting loss of brain tissue in asymptomatic individuals carrying an autosomal dominant mutation known to cause AD more than 2 years before the appearance of symptoms [15, 16].

The BBSI technique might be particularly useful for detecting disease in asymptomatic subjects at high risk of developing AD. Prospective measurements of brain atrophy have become a very relevant issue with the advent of drugs that might alter the natural history of AD. Rates of brain atrophy measured from serial registered MR are being used as a surrogate outcome of drug effectiveness in clinical trials.

In 288 AD patients participating in the phase-IIa Abeta (Aβ) immunotherapy trial, cerebral volume changes were measured from registered scan pairs at 11-month intervals [17]. Despite the reduction in the study power due to early termination owing to meningoencephalitis, significant volumetric changes were found. In particular, 45 patients who were antibody responders had greater brain volume decrease and ventricular enlargement than the placebo group (3.12 vs 2.04% and 1.10 vs 0.48%, respectively). The unexpected finding of brain volume loss was accompanied by that of a better performance on neuropsychological tests in antibody responders than in the placebo group. Volume changes might be due to amyloid removal and associated cerebral fluid shifts.

Acknowledgements. Cristina Testa, Samantha Galluzzi, and Francesca Sabattoli helped with the drafting of the manuscript.

References

1. Laakso MP, Partanen K, Riekkinen P et al (1996) Hippocampal volumes in Alzheimer's disease, Parkinson's disease with and without dementia, and in vascular dementia: an MRI study. Neurology 46:678-681
2. Laakso MP, Soininen H, Partanen K et al (1998) MRI of the hippocampus in Alzheimer's disease: sensitivity, specificity, and analysis of the incorrectly classified subjects. Neurobiol Aging 19:23-31
3. Jack CR, Jr, Petersen RC, Xu YC et al (1999) Prediction of AD with MRI-based hippocampal volume in mild cognitive impairment. Neurology 52:1397-1403
4. Jack CR, Jr, Petersen RC, Xu Y et al (2000) Rates of hippocampal atrophy correlate with change in clinical status in aging and AD. Neurology 55:484-489
5. Fox NC, Freeborough PA (1997) Brain atrophy progression measured from registered serial MRI: validation and application to Alzheimer's disease. J Magn Reson Imaging 7:1069-1075
6. Ashburner J, Csernansky JG, Davatzikos C et al (2003) Computer-assisted imaging to assess brain structure in healthy and diseased brains. Lancet Neurol 2:79-88
7. Ashburner J, Friston KJ (2000) Voxel-based morphometry: the methods. Neuroimage 14:805-821

8. Thompson PM, Mega MS, Toga AW (2000) Disease-specific brain atlases. In: Toga AW, Mazziotta JC (eds) Brain mapping: the disorders. Academic Press, San Diego, CA, pp 131-177
9. Freeborough PA, Fox NC, Kitney RI (1997) Interactive algorithms for the segmentation and quantitation of 3-D MRI brain scans. Comput Methods Programs Biomed 53:15-25
10. Chetelat G, Landeau B, Eustache F et al (2005) Using voxel-based morphometry to map the structural changes associated with rapid conversion in MCI: a longitudinal MRI study. Neuroimage 27:934-946
11. Thompson PM, Toga AW (1996) A surface-based technique for warping 3-dimensional images of the brain. IEEE Trans Med Imaging 15:1-16
12. Thompson PM, Hayashi KM, de Zubicaray G et al (2003) Dynamics of gray matter loss in Alzheimer's disease. J Neurosci 23:994-1005
13. Vidal CN, Rapoport JL, Hayashi KM et al (2006) Dynamically spreading frontal and cingulate deficits mapped in adolescents with schizophrenia. Arch Gen Psychiat 63:25-34
14. Smith SM, De Stefano N, Jenkinson M, Matthews PM (2001) Normalized accurate measurement of longitudinal brain change. J Comput Assist Tomogr 25:466-475
15. Resnick SM, Goldszal AF, Davatzikos C et al (2000) One-year age changes in MRI brain volumes in older adults. Cereb Cortex 10:464-472
16. Davatzikos C, Resnick SM (1998) Sex differences in anatomic measures of interhemispheric connectivity: correlations with cognition in women but not in men. Cereb Cortex 8:635-640
17. Fox NC, Black RS, Gilman S et al (2005) Effects of Aβ immunization (AN1792) on MRI measures of cerebral volume in Alzheimer disease. Neurology 64:1563-1572

Chapter 14

Other Neurodegenerative Conditions

M. MASCALCHI

Introduction

Neurodegenerative diseases of the central nervous system (CNS) represent a large group of sporadic or inherited conditions which share the fundamental physiopathological feature of a progressive dysfunction and death of neurons belonging to different systems, which ultimately leads to regional and global brain atrophy. Although considerable advances in understanding the etiopathogenesis of these diseases have been made, to date no treatment capable of halting the progression of neurodegeneration in any of these diseases is available.

All the non-conventional MR techniques used in patients with multiple sclerosis (MS), including volumetry, diffusion-weighted imaging (DWI) and diffusion tensor imaging (DTI), magnetization transfer imaging (MTI), proton MR spectroscopy (^{1}HMRS), and functional MR imaging (fMRI), have been also applied to the investigation of neurodegenerative diseases of the CNS.

While the technical-methodological background for their application in neurodegenerative diseases is the same as for MS, the results obtained need to take into account the invariably progressive course of neurodegenerative diseases as opposed to the typical relapsing and remitting course of the most frequent form of MS. This implies that symptomatic patients with neurodegenerative diseases are usually examined when the degeneration process has been progressing for many years, and hence the data from these patients reflect the advanced stages of neurodegeneration. An important exception is the case of subjects with genetically inherited conditions, in which MRI studies in presymptomatic carriers of the genetic mutation are feasible and have been shown to provide some insight into the early phases of neurodegeneration. Moreover, although neurodegenerative diseases of the CNS are typically system-specific, the distribution of changes is elective (but not exclusive). Accordingly, neuronal systems whose damage is not associated with clinical manifestations often also display some MR abnormalities, although these are usually less pronounced than in the elective sites of damage.

Rationale for the Use of Non-conventional MR Techniques for the Study of Degenerative Diseases of the CNS

Volumetry is an important tool in the MR armamentarium used to evaluate neurodegenerative diseases. It enables estimation of regional and global atrophy, namely the ultimate stage of neurodegeneration, and identification of specific disease markers, such as regional atrophy of the hippocampus in Alzheimer's disease, which are valuable for the diagnostic work-up.

While volumetric assessment identifies the presence and distribution of irreversible brain damage in terms of gray or white matter tissue loss, evaluation of other MR parameters, such as apparent diffusion coefficient (ADC) calculated from DWI, mean diffusivity (MD) and fractional anisotropy (FA) calculated from DTI, and magnetization transfer ratio (MTR) calculated from MTI, can estimate the integrity of the remaining brain tissue [1, 2]. For this reason, DWI, DTI, and MTI have found considerable and increasing applications in the evaluation of neurodegenerative diseases of the CNS.

^{1}HMRS provides an in vivo, non-invasive, semi-quantitative or quantitative evaluation of brain metabolites whose amounts reflect different physiopathological conditions. In particular, for neurodegenerative diseases of the CNS, the most important metabolite is N-acetylaspartate (NAA), which is a marker of neuronal density and function and, as such, is commonly decreased in these conditions [3]. In clinical practice, the amount of NAA is usually related to the amount of creatine (Cr) in the same volume of interest, which is assumed to be constant. This enables the use of the NAA/Cr ratio as a practical tool for assessment of the amount of NAA in individual patients.

fMRI exploits the changes in the oxygenation of the blood in brain areas that are activated during different tasks. Potential uses of fMRI in neurodegenerative diseases of the CNS include: detection of an altered function in the presymptomatic carriers of the mutation of a given inherited condition; exploration of the compensatory functional mechanisms operating in symptomatic patients and, in case of longitudinal studies, of their progressive failure; and monitoring the functional changes associated with new therapies.

Regional versus Whole-Brain Analysis

Two general approaches have been developed to analyze the data from non-conventional MR techniques in neurodegenerative diseases of the CNS.

The first approach is based on the search for regional differences, and is typically carried out by using a manually defined region-of-interest (ROI), usually consisting of a few pixels, or by placing the volume of interest for ^{1}HMRS in areas of expected maximal change for a certain disease. This method can easily be performed and has no complex methodological background. Its drawbacks are the degree of operator-dependence and the necessary a priori knowledge of the distribution of the abnormalities. To overcome these limitations, voxel-based analysis of MR images is used for the detection of regional changes of the brain in neurodegenerative disease of the CNS. Voxel-based analysis of MR images is a powerful tool for assessing regional structural differences in the brain between groups of subjects, without the need for any specific a priori hypothesis [4]. Originally developed as voxel-based morphometry (VBM) to investigate regional differences in gray and white matter volumes in heavily T1-weighted images (Fig. 1), the method has subsequently been utilized to investigate maps of other parameters such as the ADC, MD, FA, and MTR. Its drawbacks are the necessity to implement dedicated software packages, the complex mathematical assumptions and methodological background involved and, especially, the fact that voxel-based methods are not suited to individual subject analysis, and this limits its application in clinical practice.

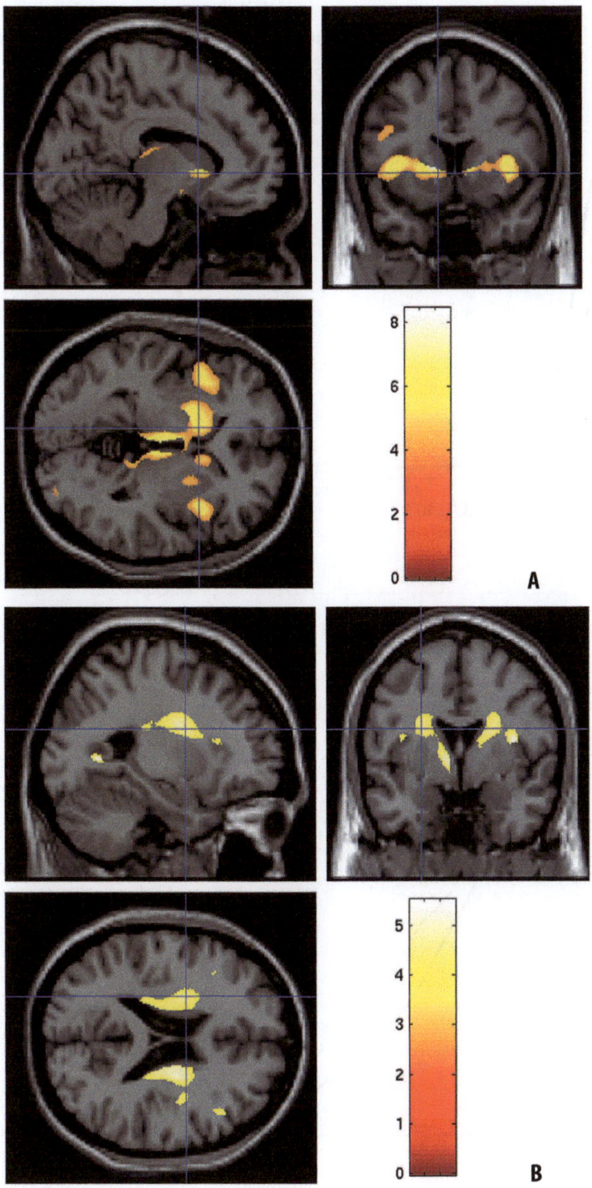

Fig. 1. Voxel-based morphometry (VBM) in patients with Huntington's disease. Statistical parametric map (SPM) representation of the multiple clusters of decreased gray matter (**A**) and white matter (**B**) volumes ($P < 0.05$ corrected for multiple comparisons) in 9 symptomatic carriers of the HD mutation compared with 11 healthy controls. The clusters are superimposed onto normalized T1-weighted images of a normal subject

The second approach consists of the assessment of a certain MR parameter for the entire brain volume [5, 6] (Fig. 2), or a metabolite concentration in ^{1}HMRS [7]. The advantages of this approach include the strong intrinsic statistical power, due to the fact that thousands of pixels are estimated as compared with the

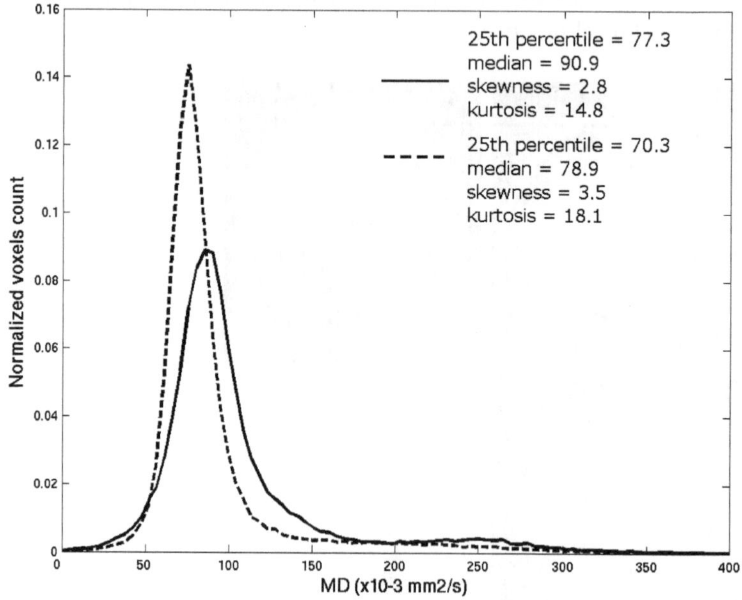

A

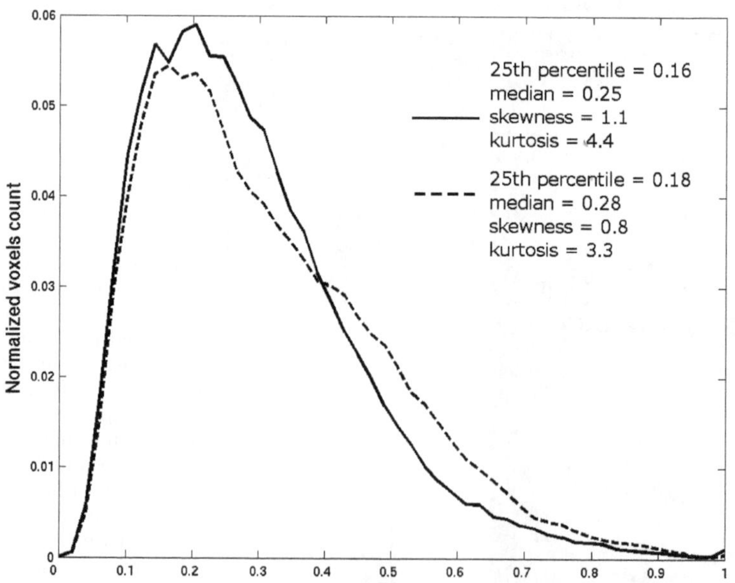

B

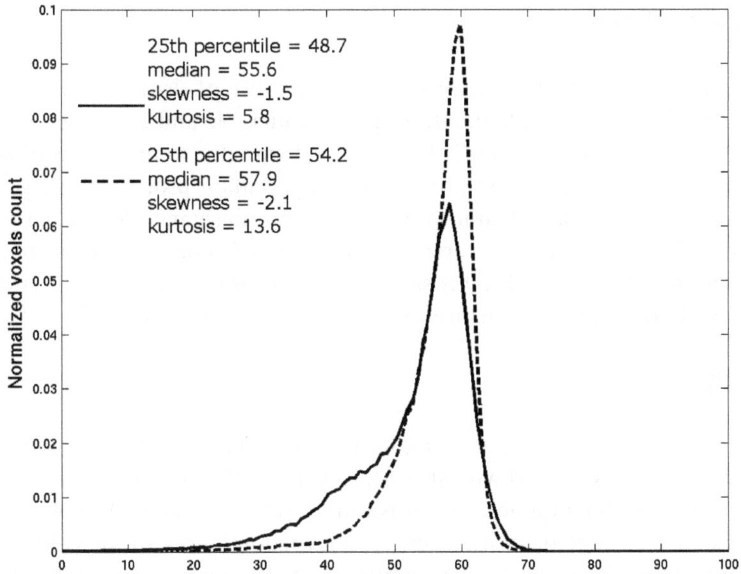

C

Fig. 2 A-C. Whole-brain histograms of the mean diffusivity (**A**), fractional anisotropy (**B**), and magnetization transfer ratio (**C**) in a patient with Huntington's disease (continuous line) and a control subject (dashed line). The patient's MD histogram (**A**) shows decreased kurtosis (curve peakedness) and skewness (asymmetry of the curve) and increased 25th and 50th (median) percentiles, compared with those of the control subject, reflecting a general increase of the MD values. Conversely, the patient's FA histogram shows increased kurtosis and skewness and decreased 25th and 50th percentiles, compared with those of the control subject, reflecting a general decrease of the FA (**B**). Finally, the patient's MTR histogram (**C**) shows decreased kurtosis, 25th and 50th percentiles, and increased skewness, reflecting a general decrease of the MTR

few usually considered for ROI analysis, and the capability to effectively provide a measurement of the global effect of the disease on the brain, which is a promising strategy for monitoring therapeutic trials. Its main drawback is the loss of spatial information.

A compromise between the ROI and whole-brain approach is to perform a histogram analysis on large volumes selected after anatomic segmentation [5].

In this chapter, the main results obtained with the use of non-conventional MR techniques in the most common neurodegenerative diseases of the CNS are reviewed. Particular attention is given to the results obtained in genetically determined conditions such as familial Alzheimer's disease, Huntington's disease, spinocerebellar ataxias, and Friedreich's ataxia, which may be viewed as interesting models to investigate the brain changes in neurodegenerative diseases of the CNS, due to the certainty of diagnosis and the feasibility of assessment from their presymptomatic stages.

Alzheimer's Disease

Alzheimer's disease (AD) is the most common cause of dementia, as well as a growing public health problem, affecting a considerable and increasing proportion of the elderly population [8]. The familial form, due to mutations of presenilin 1 (PS1), presenilin 2 (PS2), or β-amyloid precursor protein (APP) genes is a very rare cause of AD, with autosomal dominant inheritance. Molecular genetic tests enable us to identify the carriers of the mutated gene [8]. The histopathology of sporadic and familial AD is similar and is characterized by an accumulation of amyloid plaques and neurofibrillary tangles and by neuronal loss [2].

Volumetry

Volumetry of the hippocampus using ROI is an established marker for measuring tissue damage associated with sporadic AD [2, 9, 10]. In cross-sectional studies, the volume of the hippocampus was found to be reduced by 9%-60% in AD as compared with controls, with a general left predominance of the volume loss. Moreover, longitudinal studies have emphasized that there is an increased atrophy rate of the hippocampus and whole brain in AD [2]. Less well established is the capability of hippocampal volumetry to predict conversion from mild cognitive impairment to AD [2].

Several studies have investigated patients with AD, using VBM [11-13]. Baron et al. [11] compared mild AD patients and healthy age-matched controls and observed symmetric gray matter loss in the medial temporal structures, posterior cingulated gyrus, precuneus, temporoparietal association cortex, and perisylvian cortex. In a later longitudinal study, Chételat et al. [12] reported a higher rate of atrophy in the temporal neocortex (predominant on the left side) in patients with amnestic mild cognitive impairment who subsequently converted to AD. Finally, in a cross-sectional study, Frisoni et al. [13] showed greater temporoparietal atrophy in elderly patients with early-onset AD as compared to patients with late-onset AD, who conversely showed greater medial temporal atrophy. These findings were felt to be consistent with the hypothesis that "genetic factors may drive the susceptibility to developing AD lesions in the neocortex in young age, while environmental factors might exert a similar effect on medial temporal lobe structures at older age."

In familial AD, Fox and colleagues [14], using ROI analysis, demonstrated that atrophy of the hippocampus was present in the presymptomatic stage in three subjects with APP mutation who later became symptomatic. Using analysis of MR images with a voxel-compression mapping method, the same group [15] reported that temporoparietal neocortical, posterior cingulate, and medial temporal lobe atrophy were also apparent in the presymptomatic phases. Recently, the same group reported that longitudinal measures of global and hippocampal atrophy can identify differences between carriers of APP or PS1 mutations and controls 2-3 years earlier than cross-sectional measures [16].

DWI, DTI, and MTR

Almost all studies which utilized a ROI approach have documented increased ADC or MD and decreased FA in the cerebral white matter of patients with AD [2]. The analysis of the ADC in the gray matter structures has provided conflicting results, with evidence of increased ADC in the hippocampus and cingulated gyrus in some reports only [17, 18]. A correlation between MD or FA in the white matter and measurements of cognitive decline such as the Mini-Mental State Examination (MMSE) was reported [2]. In a cross-sectional study evaluating with voxel-based analysis maps of FA in elderly patients with mild sporadic AD, Xie et al. [19] found decreased white matter anisotropy in the association regions, including temporal lobes, superior longitudinal fasciculi, internal capsule and cerebral peduncles, bilaterally, and corpus callosum. Overall, the above findings are consistent with widely distributed disintegration of the white matter in AD, in agreement with a neuropathological demonstration of myelin abnormalities in the absence of axonal damage [20].

Few studies have evaluated, with ROI, the MTR of the gray and white matter in AD [2]. A decrease in the MTR in the gray matter including the hippocampus [21] was also found in patients without significant regional atrophy. This opens up the possibility that MTI might provide an early marker of neurodegeneration in the gray matter structures of the medial temporal lobe, which are the first structures involved in AD pathology.

^{1}HMRS

The neurodegenerative process in AD determines a nonspecific decrease in NAA/Cr which is combined with a relative increase in myo-inositol (mI), usually expressed by an increased mI/Cr ratio, which can be evaluated if short echo time proton spectroscopy is employed [22]. This latter feature is thought to reflect gliosis, which takes place along with neuronal loss. The distribution of the metabolic changes matches that of the AD pathology, with elective involvement of the medial temporal lobe structures, the parieto-temporal-occipital association cortex, and the cingulum [23]. ^{1}HMRS abnormalities are also observed in patients without overt atrophy [22]. Recently, Godbolt et al. [24], in a study of seven asymptomatic carriers with PS1 and APP mutations, demonstrated a decrease in the NAA/mI in the posterior cingulate cortex. The metabolic changes demonstrated by ^{1}HMRS, in particular the NAA/mI ratio, have been shown to correlate with measurements of cognitive decline and are capable of predicting the severity of the decline itself [22].

Using a whole-brain HMRS technique, Falini et al. [7] reported that the concentration of NAA distinguished patients with AD and patients with mild cognitive impairment from healthy controls.

^{1}HMRS is currently employed as a surrogate marker in treatment trials of AD. In one of these studies [25], an increase in the NAA concentration or of the

NAA/Cr was associated with treatment response, and a low baseline NAA/Cr ratio predicted a positive treatment outcome.

fMRI

During verbal retrieval tasks, patients with AD show decreased activation of the hippocampus and increased lateral prefrontal activity, the latter feature suggesting involvement of compensatory mechanisms [26]. In a study exploring verbal episodic memory [27], patients with AD showed decreased activations both of the medial temporal lobes and parieto-temporal associative areas and an increased activation of the left prefrontal cortex. In this study, a correlation of the functional pattern with the regional brain atrophy assessed with VBM was also performed. A positive correlation of the volume of the hippocampus with the activation of the medial temporal lobe and associative cortex, but not with the activation in the prefrontal cortex, was found. These data suggest that functional changes may be closely correlated to the progressive structural changes observed in the hippocampal region.

Huntington's Disease

Huntington's disease (HD) is associated with expansion of a CAG triplet in the gene IT15 of chromosome 4, a gene that encodes a protein of unknown function named huntingtin. HD is an autosomal dominant disease whose diagnosis can be established in vivo with genetic molecular testing in presymptomatic and symptomatic subjects. Clinical presentation of HD generally occurs in middle-age and includes a mixed motor disorder with involuntary movements (chorea) and reduced speed of voluntary movements (bradykinesia), variably combined with psychopathological changes and intellectual decline.

Volumetry

Using ROI analysis, loss of bulk of the neostriatum was documented in presymptomatic stages of the disease [28] and a correlation between the severity of atrophy with the number of the triplet expansion was observed [29]. The progression rate of neostriatal atrophy was analyzed in longitudinal studies, which demonstrated that the rate is not constant, ranging from a mean volume change per year of the caudate of 2.43% in presymptomatic subjects to 7.24% in patients in the advanced stages of the disease [30]. In symptomatic patients, global brain atrophy is present and, as with the neostriatal atrophy, correlates with clinical measurements of disease severity [6].

VBM results in symptomatic HD patients demonstrated a general loss of the cerebral gray matter and white matter, with a predominant gray matter loss in the striatal areas, thalamus, hypothalamus, and opercular cortex [31, 32] (Fig. 1).

Thieben et al. [33] evaluated presymptomatic carriers of the HD gene and demonstrated a volume decrease not only in the gray matter of cortical and subcortical structures, but also in the white matter of the cerebral hemispheres.

DWI, DTI, and MTR

Two studies evaluated tissue integrity in patients with HD using DWI [6, 34]. In both of them, an increased ADC was observed in the neostriatum, which was correlated with the clinical deficits and CAG repeats. In one of these studies, an increased ADC of the cerebral white matter and whole brain were also observed [6]. Similar results are obtained when the whole-brain MD and FA are measured (Fig. 2).

In a study which explored MTR in HD patients, mild nonsignificant decreases of MTR were observed in the striatum and whole brain [6] (Fig. 2).

^{1}HMRS

Single-voxel ^{1}HMRS in patients with HD was performed in cortico-subcortical regions and basal ganglia and showed a decrease in NAA in terms of increased Cr/NAA or a decrease in NAA/choline (Cho) ratios [35, 36]. An increase in lactate (Lac), which is likely to reflect an impairment of energy metabolism in HD, was also found [35, 36]. The Lac/NAA ratio was found to correlate with CAG repeats and the duration of symptoms [37], whereas the decrease in NAA/Cho correlated with the advancing clinical stage of the disease [36]. ^{1}HMRS has also been used as a surrogate marker in a pilot trial with dietary creatine therapy in HD [38].

fMRI

Functional MRI studies using non-motor tasks revealed changes in the activation of cortical and subcortical gray matter in presymptomatic HD mutation carriers [39, 40]. In particular, using a time-discrimination task, it was observed that HD mutation carriers far from clinical onset, compared with healthy controls, showed an increased activation of the medial hemispheric structures (anterior cingulated, presupplementary motor area), whereas carriers close to clinical onset, compared with controls, showed a reduced activation of the caudate and thalamus [40]. A recent study demonstrated that the pattern of activations revealed by fMRI during a simple motor task (hand tapping with the dominant hand) is somewhat different in symptomatic HD patients with respect to healthy controls, with evidence of areas of decreased activations in the caudate nucleus and medial frontal and anterior cingulate gyri, and increased activations in the supplementary motor area, supramarginal gyrus, and intraparietal sulcus [32]. In this study, fMRI findings were correlated with the distribution of gray matter loss, as measured by VBM, and a topographical correlation between areas of decreased activation and significant volume loss was found for the caudate nucleus only. This supports the notion that cortical functional changes in HD are likely to reflect decreased output of the motor basal ganglia-thalamo-cortical circuit, with compensatory recruitment of additional accessory motor pathways.

Parkinson's Disease

Volumetry

VBM studies have investigated PD patients in relatively advanced stages of the disease [41-44] and showed extensive clusters of gray matter volume reduction in the frontal and limbic-paralimbic regions. In PD patients with dementia, the volume loss extended to temporal, parietal, and occipital lobes as well as to thalamus and caudate. Notably, no VBM study showed midbrain changes in PD, whereas such changes were observed in patients with progressive supranuclear palsy [45].

DWI, DTI, and MTR

Only a few studies have investigated MTR of the brain in patients with PD; they all used ROI analysis and achieved conflicting results. In one study [46], PD patients with dementia had reduced MTR values in the subcortical white matter, but there were no significant differences in any region between non-demented PD patients and controls. In another study [47], lower MTR in supratentorial white matter and brainstem was found in PD patients without dementia as compared with controls. More recently, Eckert et al. [48] reported reduced MTR in the globus pallidus, putamen, caudate nucleus, substantia nigra, and white matter in PD.

Also, most MR studies of diffusion in PD patients adopted ROI analysis and focused on subcortical nuclei, mainly to define features useful for differential diagnosis with other parkinsonisms [49, 50]. In these studies, no significant differences in ADC values in the subcortical gray matter between PD patients and controls were found. On the other hand, in another ROI study, Yoshikawa et al. [51] demonstrated that patients with PD have decreased FA in the white matter corresponding to the nigrostriatal projections, and that a significantly reduced FA in subcortical frontal white matter of the premotor area occurred in the most advanced stages of the disease. Finally, in a recent study which used voxel-based analysis of the trace of the ADC [52], an increased diffusivity of the olfactory tracts in PD patients was found, and this has been proposed as a possible early diagnostic marker of PD.

^{1}HMRS

Bowen et al. [53] demonstrated a significant increase in the lactate/Cr ratio in the occipital lobe of patients with PD, which was more pronounced in patients with dementia; whereas Hu et al. [54] reported temporoparietal cortical reductions in NAA/Cr ratios in non-demented PD patients.

fMRI

Using a motor task (single joystick movements), Haslinger et al. [55] demonstrated, in patients with akinetic PD both before and 30 minutes after taking oral levodopa, an impaired activation of the supplementary motor area and increased activations of the primary motor cortex and the lateral premotor cortex, bilaterally, compared with healthy controls. Levodopa administration was associated with a relative normalization of the activation pattern.

Inherited Ataxias: Spinocerebellar Ataxias Type 1 and 2

Volumetry

SCA1 and SCA2 are relatively common autosomal dominant adult-onset ataxias which have a neuropathological pattern similar to that of olivopontocerebellar atrophy.

Patients with SCA1 have a decrease in the volumes of the cerebellum and brainstem without atrophy of the basal ganglia and cerebral hemispheres [56, 57]. In patients with SCA2, atrophy of the cerebellum and brainstem is generally more marked than in patients with SCA1 [56-58]. A VBM study of SCA2 showed evidence of cerebral cortical atrophy [58]. A correlation between severity of the clinical deficit and atrophy of the brainstem was observed in SCA1, but not in SCA2 [57].

ADC, DTI, and MTR

Increases in ADC values of the brainstem and cerebellum with normal ADC of the cerebral hemispheres have been reported in SCA1 and SCA2 [57]. In the same study, a correlation between the clinical severity and the increase in the brainstem and cerebellar ADC values was observed in SCA1, but not in SCA2.

^{1}HMRS

^{1}HMRS shows a decrease in the NAA/Cr ratio, NAA concentration, and Cho/Cr ratio in the pons and the deep cerebellum in asymptomatic and symptomatic carriers of the SCA1 mutation [57, 59].

In patients with SCA2, a decrease in NAA/Cr ratio and NAA concentration in the pons and cerebellum were reported [57, 60]. In one study, an increase in the cerebellar Lac was noted [60], whereas in another study an increase in the glial marker mI was documented in the pons and cerebellum [61]. A correlation between clinical severity and the decrease in the NAA/Cr ratio in the pons was observed in SCA1, but not in SCA2 [57].

fMRI

During a simple motor task (hand tapping with the dominant hand), patients with SCA1 and SCA2 showed an increased activation of the contralateral premotor cortex and a decreased activation of the normal cerebellar activation, as compared with healthy controls (unpublished personal observations) (Fig. 3).

Inherited Ataxias: Spinocerebellar Ataxia Type 6

SCA6 is an autosomal dominant adult-onset cerebellar ataxia which shares with idiopathic late-onset pure cerebellar ataxia (ILOCA) a neuropathological pattern of cortical cerebellar atrophy.

Volumetry

In a volumetric MRI study of ILOCA and SCA6 [62], an inverse correlation was found between the total cerebellar volume and the severity of the cerebellar deficits, as well as between the regional volumes of the cerebellum and partial scores of the corresponding functional system.

ADC, DTI, and MTR

The ADC was significantly increased in the cerebellum and brainstem of patients with ILOCA as compared with healthy controls, possibly because of mild structural damage to the cerebellopetal and cerebellofugal white matter tracts [63]. In the same study, the ADC and volume of the cerebral hemispheres were not significantly different between patients and controls.

^{1}HMRS

A decrease in NAA/Cr without increased Lac was observed in the cerebellum, pons, and frontal cortex of patients with SCA6 or ILOCA [60, 64, 65].

Inherited Ataxias: Friedreich's Ataxia

Friedreich's ataxia (FRDA) is an autosomal recessive disease with an onset in childhood or adolescence. At the neuropathological examination, FRDA exhibits a pattern of spinal atrophy.

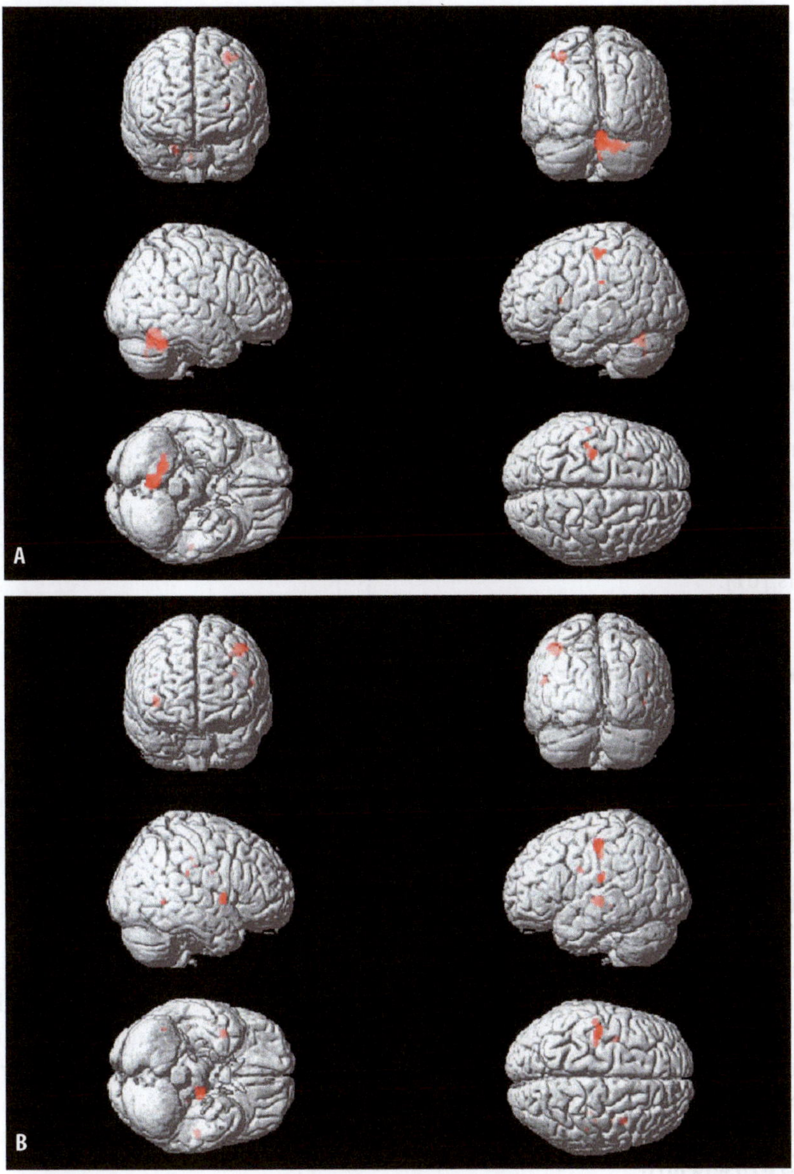

Fig. 3 A, B. Functional MRI during hand tapping of the dominant hand in 18 healthy controls (**A**) and 6 patients with SCA1 and 10 patients with SCA2 (**B**). SPM maps color-coded for t values superimposed onto high resolution T1-weighted images (normal templates) showing the within-group comparison between activation and rest. Clusters of voxels in the resulting Z statistical images with a height threshold of $P < 0.001$ (uncorrected) and an extent threshold of $P < 0.05$ (corrected) were considered as significant. Note the normal activation of the primary motor cortex and contralateral cerebellum in the healthy subjects (**A**), and the activation of the contralateral premotor cortex and decreased cerebellar activation in patients with SCA1 and SCA2 (**B**)

Volumetry

In patients with FRDA, MRI studies show marked atrophy of the spinal cord and medulla, whereas the loss of bulk of the cerebellum is usually not remarkable and there is no atrophy of the cerebral hemispheres [63].

ADC, DTI, and MTR

In line with neuropathological and conventional MRI findings, the ADC was found to be reduced only in the medulla of patients with FRDA [63].

^{1}HMRS

A mild, but significant, decrease in the NAA/Cr with normal Cho/Cr and mI/Cr was observed in the pons and deep cerebellar hemisphere [64] in patients with FRDA. Another study reported a decrease in the concentration of Cho in the cerebellar vermis of FRDA patients, compared with healthy controls [61].

Motor Neuron Disease

Motor neuron disease (MND) is a term comprising different conditions characterized by selective degeneration of the upper (primary lateral sclerosis) or lower (progressive spinal muscular atrophy) motor neurons or both (amyotrophic lateral sclerosis – ALS), whose speed of progression is variable.

Volumetry

Whole-brain atrophy is not a feature of MND, but a localized decrease of the gray matter volume in the superior, middle, and medial frontal gyri, bilaterally, and of the white matter from the precentral sulcus to the brainstem, consistent with the location of the corticospinal tract, was found [66]. Cervical cord atrophy may be apparent, especially in patients with ALS [67].

ADC, DTI, and MTR

DWI and DTI enable us to investigate the degenerating corticospinal tract in the brain of patients with involvement of the upper motor neuron, revealing increased ADC and MD and reduced FA, typically in the posterior limb of the internal capsule [68-70]. These changes, which can be detected before clinical manifestation of upper motor neuron involvement, are correlated with measures of disease severity and duration, can support a diagnosis, and may allow us to monitor the evolution of the disease in individual patients. A decrease of MTR was observed in the degenerating corticospinal tract of patients with upper motor neuron involvement [71].

Recently, FA decrease has also been documented in the cervical spinal cord of patients with ALS, and this was found to be strictly associated with clinical disability [67].

^{1}HMRS

^{1}HMRS with voxels of interest placed in the motor cortex or in the brainstem demonstrates a decrease in the NAA/Cr ratio in patients with upper motor neuron involvement, which may be useful in the diagnostic work-up of individual patients [72]. Preliminary data indicating reversal of these metabolite changes after treatment with riluzole in MND [73] were not confirmed.

fMRI

In a study where the force exerted by the patients in a finger flexion task was strictly controlled, Konrad et al. [74] found that patients with amyotrophic lateral sclerosis showed a more anterior location of the motor cortex activation cluster, an increased activation of the supplementary motor area, which was also shifted anteriorly, and an increased activation of the premotor and parietal cortex, bilaterally.

Conclusions

Non-conventional MR techniques have an important role to play in the assessment of neurodegenerative diseases of the CNS. In particular, they can contribute to diagnosis in individual patients and provide important insights into the pathophysiology of these conditions.

Finally, they may also constitute reliable tools for monitoring the progression of neurodegeneration and the efficacy of new therapies. The role of MRI data as "surrogate markers" in clinical trials is justified by the insensitivity of the clinical assessment. However, in view of future multi-center trials based on such imaging modalities as "surrogate markers," standardization and optimization of imaging protocols are mandatory and should be encouraged.

Acknowledgements. The generous support of the National Organization for Rare Disorders (Danbury, CT, USA) is acknowledged.

References

1. Mascalchi M, Filippi M, Floris R et al (2005) Diffusion-weighted MR of the brain: methodology and clinical applications. Radiol Med (Turin) 109:155-197
2. Ramani A, Jensen JH, Helpern JA (2006) Quantitative MR imaging in Alzheimer's disease. Radiology 241:26-43

3. Gill SS, Small RK, Thomas DG et al (1989) Brain metabolites as ^{1}HNMR markers of neuronal and glial disorders. NMR Biomed 2:196-200
4. Ashburner J, Csernansky JG, Davatzikos C et al (2003) Computer-assisted imaging to assess brain structure in healthy and diseased brain. Lancet Neurol 2:79-88
5. Bozzali M, Franceschi M, Falini A et al (2001) Quantification of tissue damage in AD using diffusion tensor and magnetization transfer MRI. Neurology 57:1135-1137
6. Mascalchi M, Lolli F, Della Nave R et al (2004) Volumetric, diffusion and magnetization transfer MR imaging of the brain in individuals with Huntington's disease. Radiology 232:867-873
7. Falini A, Bozzali M, Magnani G et al (2005) A whole brain MR spectroscopy study from patients with Alzheimer's disease and mild cognitive impairment. NeuroImage 26:1159-1163
8. Blennow K, de Leon M, Zetterberg H (2006) Alzheimer's disease. Lancet Neurol 368:387-403
9. Jack CR, Shiung MM, Gunter JL et al (2004) Comparison of different MRI brain atrophy rate measures with clinical disease progression in AD. Neurology 62:591-600
10. Testa C, Laakso MP, Sabattoli F et al (2004) A comparison between the accuracy of voxel-based morphometry and hippocampal volumetry in Alzheimer's disease. J Magn Reson Imaging 19:274-282
11. Baron JC, Chételat G, Desgranges B et al (2001) In vivo mapping of gray matter loss with voxel-based morphometry in mild Alzheimer's disease. NeuroImage 14:298-309
12. Chételat G, Landeau B, Eustache F et al (2005) Using voxel-based morphometry to map the structural changes associated with rapid conversion in MCI: a longitudinal MRI study. NeuroImage 27:934-946
13. Frisoni G, Testa C, Sabattoli F et al (2005) Structural correlates of early and late onset Alzheimer's disease: voxel based morphometry study. J Neurol Neurosurg Psychiat 76:112-114
14. Fox NC, Warrington EK, Freeborough PA, Hartikainen P, Kennedy AM, Stevens JM, Rossor MN (1996) Presymptomatic hippocampal atrophy in Alzheimer's disease: a longitudinal MRI study. Brain 119:2001-2007
15. Fox NC, Crum WR, Scahill RI, Stevens JM, Janssen J, Rossor M (2001) Imaging of onset and progression of Alzheimer's disease with voxel-compression mapping of serial magnetic resonance images. Lancet 358:201-205
16. Ridha BH, Barnes J, Bartlett JW, Godbolt A, Pepple T, Rossor MN, Fox NC (2006) Tracking atrophy progression in familial Alzheimer's disease: a serial MRI study. Lancet Neurol 5:828-834
17. Kantarci K, Jack CR, Yu YC et al (2001) Mild cognitive impairment and Alzheimer disease: regional diffusivity of water. Radiology 219:101-107
18. Sandson TA, Felician O, Edelman RR, Warach S (1999) Diffusion-weighted magnetic resonance imaging in Alzheimer's disease. Dementia Geriatr Cogn Disord 10:166-171
19. Xie S, Xiao JX, Gong GL et al (2006) Voxel-based detection of white matter abnormalities in mild Alzheimer disease. Neurology 66:1845-1849
20. Umahara T, Tsuchiya K, Ikeda K et al (2002) Demonstration and distribution of tau-positive glial coiled body-like structures in white matter and white matter threads in early onset Alzheimer's disease. Neuropathology 22:9-12
21. Hanyu H, Asano T, Iwamoto T et al (2000) Magnetization transfer measurements of the hippocampus in patients with Alzheimer's disease, vascular dementia, and other types of dementia. AJNR Am J Neuroradiol 21:1235-1242
22. Valenzuela MJ, Sachdev P (2001) Magnetic resonance spectroscopy in AD. Neurology 56:592-598
23. Block W, Jessen F, Traber F et al (2002) Regional N-acetylaspartate reduction in the hippocampus detected with fast proton magnetic resonance spectroscopic imaging in patients with Alzheimer disease. Arch Neurol 59:828-834

24. Godbolt AK, Waldman AD, MacManus DG et al (2006) MRS shows abnormalities before symptoms in familial Alzheimer disease. Neurology 66:718-722
25. Jessen F, Traeber F, Freymann K et al (2006) Treatment monitoring and response prediction with proton MR spectroscopy in AD. Neurology 67:528-530
26. Backman L, Andersson JL, Nyberg L et al (1999) Brain regions associated with episodic retrieval in normal aging and Alzheimer's disease. Neurology 52:1861-1870
27. Rémy F, Mirrashed F, Campbell B, Richter W (2005) Verbal episodic memory impairment in Alzheimer's disease: a combined structural and functional MRI study. NeuroImage 25:253-266
28. Aylward EH, Brandt J, Codori AM et al (1994) Reduced basal ganglia volume associated with the gene for Huntington's disease in asymptomatic at-risk persons. Neurology 44:823-828
29. Rosas HD, Goodman J, Chen YI et al (2001) Striatal volume loss in HD as measured by MRI and the influence of CAG repeat. Neurology 57:1025-1028
30. Aylward EH, Codori AM, Rosenblatt A et al (2000) Rate of caudate atrophy in presymptomatic and symptomatic stages of Huntington's disease. Mov Disord 15:552-560
31. Kassubek J, Juengling FD, Kioschies T et al (2004) Topography of cerebral atrophy in early Huntington's disease: a voxel based morphometric MRI study. J Neurol Neurosurg Psychiat 75:213-220
32. Gavazzi C, Della Nave R, Petralli R et al (2007) Combining functional and structural MR imaging of the brain in Huntington disease. J Comput Assist Tomogr (*in press*)
33. Thieben MJ, Duggins AJ, Good CD et al (2002) The distribution of structural neuropathology in pre-clinical Huntington's disease. Brain 125:1815-1828
34. Seppi K, Schocke MF, Mair KJ et al (2006) Diffusion-weighted imaging in Huntington's disease. Mov Disord 21(7):1043-1047
35. Jenkins B, Koroshetz W, Beal MF, Rosen B (1993) Evidence for an energy metabolism defect in Huntington's disease using localized proton spectroscopy. Neurology 43:2689-2695.
36. Harms L, Meierkord H, Timm G et al (1997) Decreased N-acetylaspartate/choline ratio and increased lactate in the frontal lobe of patients with Huntington's disease: a proton magnetic resonance spectroscopy study. J Neurol Neurosurg Psychiat 62:27-30
37. Jenkins BG, Rosas HD, Chen CI et al (1998) ^{1}HNMR spectroscopy studies of Huntington's disease: correlations with CAG repeat numbers. Neurology 50:1357-1365
38. Tabrizi SJ, Blamire AM, Manners DN et al (2003) Creatine therapy for Huntington's disease: clinical and MRS findings in a 1-year pilot study. Neurology 61:141-142
39. Reading SAJ, Dziorny AC, Peroutka LA et al (2004) Functional brain changes in presymptomatic Huntington's disease. Ann Neurol 55:879-883
40. Paulsen JS, Zimbelman JL, Hinton SC et al (2004) fMRI biomarker of early neuronal dysfunction in presymptomatic Huntington's disease. AJNR Am J Neuroradiol 25:1715-1721
41. Burton EJ, McKeith IG, Burn DJ et al (2004) Cerebral atrophy in Parkinson's disease with and without dementia: a comparison with Alzheimer's disease, dementia with Lewy bodies and controls. Brain 127:791-800
42. Ramírez-Ruiz B, Martí MJ, Tolosa E et al (2005) Longitudinal evaluation of cerebral morphological changes in Parkinson's disease with and without dementia. J Neurol 252:1345-1352
43. Nagano-Saito A, Washimi Y, Arahata Y et al (2005) Cerebral atrophy and its relation to cognitive impairment in Parkinson disease. Neurology 64:224-229
44. Summerfield C, Junqué C, Tolosa E et al (2005) Structural brain changes in Parkinson disease with dementia: a voxel-based morphometry study. Arch Neurol 62:281-285

45. Price S, Paviour D, Scahill R et al (2004) Voxel-based morphometry detects patterns of atrophy that help differentiate progressive supranuclear palsy and Parkinson's disease. NeuroImage 23:663-669
46. Hanyu H, Asano T, Sakurai H et al (2001) Magnetisation transfer measurements of the subcortical grey and white matter in Parkinson's disease with and without dementia and in progressive supranuclear palsy. Neuroradiology 43:542-546
47. Tambasco N, Pelliccioli GP, Chiarini P et al (2003) Magnetization transfer changes of gray and white matter in Parkinson's disease. Neuroradiology 45:224-230
48. Eckert T, Sailer M, Kaufmann J et al (2004) Differentiation of idiopathic Parkinson's disease, multiple system atrophy, progressive supranuclear palsy and healthy controls using magnetization transfer imaging. NeuroImage 21:229-235
49. Seppi K, Schocke MF, Esterhammer R et al (2003) Diffusion-weighted imaging discriminates progressive supranuclear palsy from PD, but not from the Parkinson variant of multiple system atrophy. Neurology 60:922-927
50. Schocke MF, Seppi K, Esterhammer R et al (2004) Trace of diffusion tensor differentiates the Parkinson variant of multiple system atrophy and Parkinson's disease. NeuroImage 21:1443-1451
51. Yoshikawa K, Nakata Y, Yamada K, Nakagawa M (2004) Early pathological changes in the parkinsonian brain demonstrated by diffusion tensor MRI. J Neurol Neurosurg Psychiat 75:481-484
52. Scherfler C, Schocke MF, Seppi K et al (2006) Voxel-wise analysis of diffusion weighted imaging reveals disruption of the olfactory tract in Parkinson's disease. Brain 129:538-542
53. Bowen BC, Block RE, Sanchez-Ramos J et al (1995) Proton MR spectroscopy of the brain in 14 patients with Parkinson disease. AJNR Am J Neuroradiol 16:61-68
54. Hu MT, Taylor-Robinson SD, Chaudhuri KR et al (2000) Cortical dysfunction in non-demented Parkinson's disease patients: a combined ^{31}P-MRS and ^{18}FDG-PET study. Brain 123:340-352
55. Haslinger B, Erhard P, Kampfe N et al (2001) Event-related functional magnetic resonance imaging in Parkinson's disease before and after levodopa. Brain 124:558-570
56. Klockgether T, Skalej M, Wedekind D et al (1998) Autosomal dominant cerebellar ataxia type I: MRI-based volumetry of posterior fossa structures and basal ganglia in spinocerebellar ataxia types 1, 2 and 3. Brain 121:1678-1693
57. Guerrini L, Lolli F, Ginestroni A et al (2004) Brainstem neurodegeneration correlates with clinical dysfunction in SCA1 but not in SCA2: a volumetric, diffusion and quantitative proton spectroscopy MR study. Brain 127:1785-1795
58. Brenneis C, Bosch SM, Schocke M et al (2003) Atrophy pattern in SCA2 determined by voxel-based morphometry. Neuroreport 14:1799-1802
59. Mascalchi M, Tosetti M, Plasmati R et al (1998) Proton magnetic resonance spectroscopy in an Italian family with spinocerebellar ataxia type 1. Ann Neurol 43:244-252
60. Boesch SM, Schoecke M, Burk K et al (2001) Proton magnetic resonance spectroscopic imaging reveals differences in spinocerebellar ataxia types 2 and 6. J Magn Reson Imaging 13:553-539
61. Viau M, Machand L, Bard C, Boulanger Y (2005) ^{1}H magnetic resonance spectroscopy of autosomal ataxias. Brain Res 1049:191-202
62. Richter S, Dimitrova A, Maschke M et al (2005) Degree of cerebellar ataxia correlates with three-dimensional MRI-based cerebellar volume in pure cerebellar degeneration. Eur Neurol 54:23-27
63. Della Nave R, Foresti S, Tessa C et al (2004) ADC mapping of neurodegeneration in the brainstem and cerebellum in patients with progressive ataxias. NeuroImage 22:698-705

64. Mascalchi M, Cosottini M, Lolli F et al (2002) Proton MR spectroscopy of the cerebellum and pons in patients with degenerative ataxia. Radiology 223:371-378
65. Terakawa H, Abe K, Watanabe Y et al (1999) Proton magnetic resonance spectroscopy (1H MRS) in patients with sporadic cerebellar degeneration. J Neuroimaging 9:72-77
66. Ellis CM, Suckling J, Amaro E Jr, et al (2001) Volumetric analysis reveals corticospinal tract degeneration and extramotor involvement in ALS. Neurology 57:1571-1578
67. Valsasina P, Agosta F, Benedetti B et al (2006) Diffusion anisotropy of the cervical cord is strictly associated with disability in ALS. J Neurol Neurosurg Psychiat [epub 9 Oct 2006; doi:10.1136/jnnp.2006.100032]
68. Ellis CM, Simmons A, Jones DK et al (1999) Diffusion tensor MRI assesses corticospinal tract damage in ALS. Neurology 53:1051-1058
69. Sach M, Winkler G, Glauche V et al (2004) Diffusion tensor MRI of early upper motor neuron involvement in amyotrophic lateral sclerosis. Brain 127:340-350
70. Ulug AM, Grunewald T, Lin MT et al (2004) Diffusion tensor imaging in the diagnosis of primary lateral sclerosis. J Magn Reson Imaging 19:34-39
71. Tanabe JL, Vermathen M, Miller R et al (1998) Reduced MTR in the corticospinal tract and normal T2 in amyotrophic lateral sclerosis. Magn Reson Imaging 16:1163-1169
72. Chan S, Shungu DC, Akinwande D et al (1999) Motor neuron diseases: comparison of single voxel-proton MR spectroscopy of the motor cortex with MR imaging of the brain. Radiology 212:763-769
73. Kalra S, Cashman NR, Genge A, Arnold DL (1998) Recovery of N-acetylaspartate in corticomotor neurons of patients with ALS after riluzole therapy. Neuroreport 9:1757-1761
74. Konrad C, Henningsen H, Bremer J et al (2002) Pattern of cortical reorganization in amyotrophic lateral sclerosis: a functional magnetic resonance imaging study. Exp Brain Res 143:51-56

64. Mascalchi M, Cosottini M, Lolli F et al (2002) Proton MR spectroscopy of the cerebellum and pons in patients with degenerative ataxia. Radiology 223:371–378
65. [illegible] H, Abe K, Watanabe Y et al (1999) Proton magnetic resonance spectroscopy (1H MRS) in patients with spinocerebellar degeneration. J Neuroimaging 9:72–[illegible]
66. [illegible] Volumetric [illegible] in ALS [illegible]
67. [illegible] (2000) [illegible] Nature Neuroscience [illegible]
68. [illegible]
69. [illegible]
70. [illegible]
71. [illegible]
72. [illegible]
73. [illegible]
74. [illegible]

Chapter 15

Incorporation of Other Biomarkers

S. Gnanapavan, G. Giovannoni

Introduction

Ramòn y Cajal, considered by many to be the father of neuroscience, was the first to document in any great detail the complex structure and function of the human nervous system. In recognition of his achievements he was awarded the Nobel Prize in 1906, together with his contemporary, Camillo Golgi. Successive generations of neuroscientists added to his work, broadening our understanding of this field, but it was not until the Human Genome Project that it became possible to assimilate vast quantities of information from an entire genome or proteome. As a consequence, the number of biomolecules relevant to the study of neuronal biology and pathology has escalated. It therefore comes as no surprise to find that the literature on biomarkers is vast and encompasses a wide spectrum of disciplines from genetics to medical physics. This chapter focuses on the use of molecular biomarkers in neurodegeneration and the progress made so far.

Modeling of Neurodegeneration using Biological Markers

Neurodegeneration is the progressive damage or death of neurons, resulting in irreversible loss of function in the affected parts of the central nervous system. The pathological characteristics of neurodegeneration can be found, in varying degrees, in a number of unrelated neurological disorders, for example multiple sclerosis (MS) [1], as well as in the primary neurodegenerative disorders. Often there is a long subclinical phase before clinical symptoms become evident, and it would be ideal if markers were positive during these preclinical stages. By using markers directly related to the pathogenesis of neurodegeneration, for example biological markers of axonal injury and neurodestruction, and oxidative stress and apoptosis (Fig. 1), it may be possible to detect incipient disease with high accuracy very early in its course. Such biomarkers could also be useful in predicting prognosis and evaluating the efficacy of new therapies.

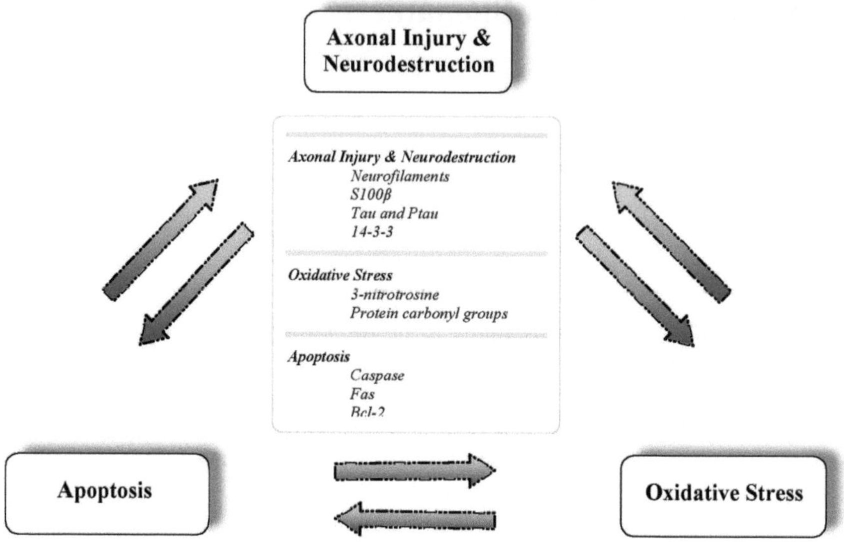

Fig. 1. Modeling of neurodegeneration using biological markers

Biological Markers of Axonal Injury and Acute Neurodestruction

Neurofilament Proteins

Neurofilaments (Nf) belong to the intermediate filament family and form the dominant cytoskeletal element in neurons, especially within the axons. They are grouped according to molecular weight into light (NfL; 70 kDa), medium (NfM; 150 kDa), and heavy chains (NfH; 200 kDa) [2]. Their function is largely structural, maintaining axonal caliber and integrity.

Neurofilament breakdown is a prominent pathogenic feature of many neurodegenerative disorders [for reviews, see 3, 4]. In sporadic and familial amyotrophic lateral sclerosis (ALS), an accumulation of neurofilaments has been observed early on in the disease using immunohistochemistry [5-7] (Fig. 2). Other neurodegenerative disorders where abnormal neurofilamentous accumulations have been observed include Alzheimer's disease (AD), Parkinson's disease (PD), progressive supranuclear palsy (PSP), other non-Alzheimer type degenerative dementias (Pick's disease, diffuse Lewy body disease (DLB), corticobasal degeneration), and dementia with neurofilament inclusions [8-10].

During neuronal breakdown, neurofilaments also leak into the adjacent cerebrospinal fluid (CSF) compartment, where they can be measured directly. Elevated CSF levels of both NfH and NfL have been found in ALS, particularly in the upper motor neuron-dominant subtype and rapidly progressive group, as well as in AD

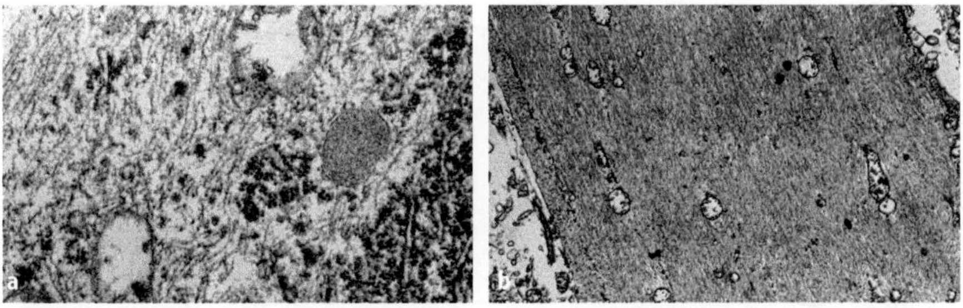

Fig. 2. Increased 10-nm neurofilaments (*wavy pattern of fine fibrils*) in the perikaryon of an anterior horn cell in ALS, ×38,000 (**a**). Longitudinal section of an enlarged neuronal process containing compactly arranged parallel neurofilaments, ×12,000 (**b**). Reproduced with permission from [6, 7]

and a variety of other neurodegenerative diseases including cerebral infarction, vascular dementia (VaD), MS, olivopontocerebellar atrophy, and normal pressure hydrocephalus [11-13]. Comparison with magnetic resonance imaging (MRI) has shown an association between CSF NfL levels and severe white matter changes, again probably representing axonal degeneration in those patients [14] (Fig. 3).

In MS, transected axons are a common feature within demyelinated plaques. The density of axons in plaques and normal-appearing white matter (NAWM) of

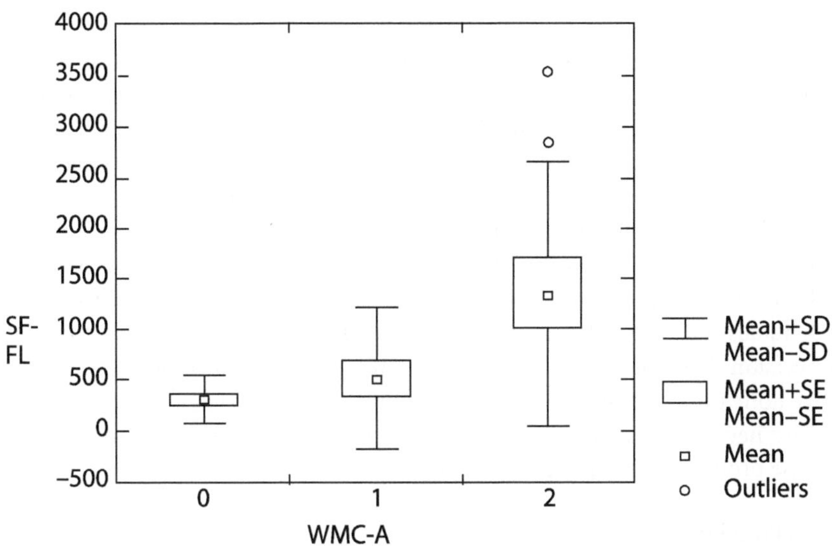

Fig. 3. Scatter plot of CSF NfL (pg/ml) in individuals with an average extent of white matter changes (WMC-A) score graded by the Blennow-Wallin scale as 0 = no lesions; 1 = only mild, punctuate lesions; and 2 = beginning of confluent lesions, to severe extensive lesions, demonstrating increasing CSF NfL levels with increasing degree of WMC. Reproduced with permission from [14]

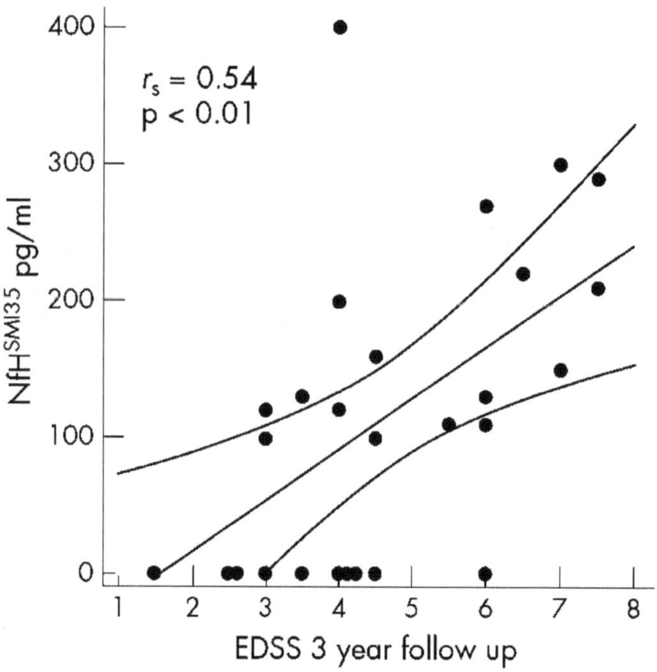

Fig. 4. Correlation between CSF NfH_{SMI135} levels and EDSS score at follow-up. Reproduced with permission from [18]

MS patients has also been shown to be significantly reduced [15, 16]. CSF NfL levels have been shown to correlate with relapses and inflammatory cell counts, as well as with disease progression [17]. More extensive studies using NfH show that baseline NfH levels predict disease progression, correlating well with various disability scores including the Expanded Disability Status Scale (EDSS) [18] (Fig. 4), the Multiple Sclerosis Functional Composite (MSFC) [18], and the recently formulated Multiple Sclerosis Severity Score (MSSS) [19]. NfH are also a prognostic indicator for recovery from optic neuritis, in which visual acuity recovers more in individuals with lower levels of NfH [20], and they can also be used to predict conversion from clinically isolated syndrome (CIS) to MS (more sensitive than MRI alone (40% vs 34%), 60% when combined with MRI) [21].

The neurofilament proteins are therefore valuable prognostic indicators of disease progression in neurodegeneration.

S-100 Proteins

S-100 proteins are a group of calcium-binding proteins involved in a multitude of cellular processes including metabolism, cell differentiation, and cell growth. The homodimer S100B (or S100β) is found at high levels in the central nervous system. Originally thought to be a marker of acute neurodestruction, it is now believed to be a marker of astroglial activation. High levels have been found in AD, Down syndrome (DS), MS, brain ischemia, and Creutzfeldt-Jakob disease (CJD) [22-25].

There is considerable overlap between the disorders, but the highest S100 levels have been found in patients with dementing illnesses. In AD [26, 27], the levels are higher in patients with mild to moderate dementia than in those with advanced dementia [28], and these levels are not influenced by age of onset or duration of disease [29]. Conversely, raised levels are found at all stages of CJD, with higher levels being associated with a shorter survival time [23]. In ALS, S100β levels have been shown to decrease with the progression of disease [30, 31].

S100B is also raised during the relapsing phase of MS [32-34], and this finding is supported by post-mortem examination findings in acute plaques [24]. No change in S100β levels has been noted during other stages of disease or in response to immunomodulatory therapy with interferon β (IFNβ)-1a [35], and no correlations were found with either the Expanded Disability Status Scale (EDSS) or MRI measures of disease activity [35].

Finally, the role of S100B in brain ischemia warrants a special mention, due to the acute nature of brain injury. S100β levels are raised after an ischemic event and reach their maximum 2-3 days after the event (Fig. 5). Levels are higher in patients

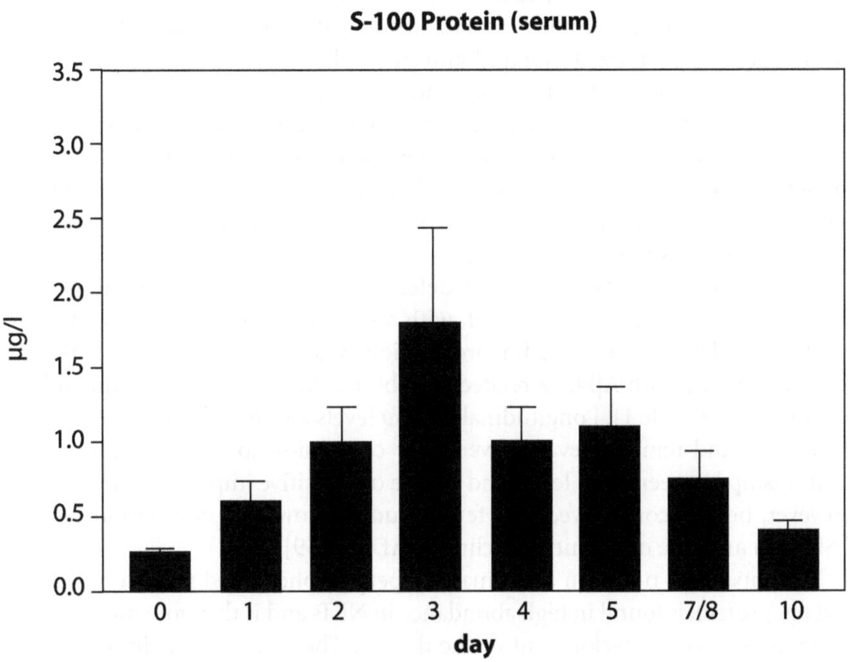

Fig. 5. Temporal profile of S100β levels in patients with acute middle cerebral artery (MCA) territory infarction. Reproduced with permission from [36]

with severe neurological deficits at the outset, and in those who have extensive infarcts, serial measurements help to predict infarct size ($r = 0.75$, $P < 0.001$) and, most importantly, neurological outcome [36-38].

Overall, S100B is an objective marker of disease progression in neurodegenerative disorders, but is not diagnostically useful in differentiating between different neurological disorders.

Tau and Phosphorylated Tau Proteins

Tau is a member of the microtubule-associated protein (MAP) family and is situated in the cytoskeleton and intracellular transport systems of neurons. Tau is released into the extracellular space through damage and degeneration of neurons. Total CSF tau levels have been found to be consistently elevated in AD pathology. Phosphorylated tau (p-tau) is a marker of neurofibrillary tangle (NFT) formation and, unlike total tau (t-tau), is a direct measure of AD pathophysiology and therefore more specific for AD. The published literature on CSF tau biomarkers is vast; detailed reviews can be found in articles by Hampel and Blennow [39-42].

Across the studies, the sensitivity and specificity of t-tau for AD was approximately 83% and 82%, respectively, using the "Innogenetics" ELISA [39]. Importantly, t-tau levels are normal in major depression, alcoholic dementia, and PD [43-45]. However, when comparing AD with other dementing disorders, the discriminatory value of t-tau was found to be low [46-48]. Moreover, in disorders characterized by rapid and/or marked neuronal destruction (CJD, acute stroke, traumatic brain injury, and early stages of RRMS), t-tau was also found to be significantly elevated [49-52]. Therefore, although t-tau is abnormal in the majority of cases, its value as a diagnostic marker for AD is limited by its inability to differentiate AD convincingly from other relevant primary dementias and neurological disorders which feature neuronal loss or destruction as part of their neuropathophysiology.

T-tau as a prognostic marker in detecting incipient AD in mild cognitive impairment (MCI) cases is excellent, with a sensitivity of 90% and specificity of 100% [53]. The hazard ratio for progression was quoted at 17.7 ($P < 0.0001$) when combined with Aβ42, a related CSF biomarker (see below) in one follow-up study [54] (Table 1). Longitudinally, t-tau levels are elevated in the CSF of cases of MCI and remain elevated even after conversion to clinical AD [55]. The relationship between t-tau levels and degree of cognitive impairment in AD have, however, been inconclusive, with fewer studies showing a correlation between CSF t-tau and rate of cognitive decline in AD [56-59].

In comparison, p-tau, an abnormally hyperphosphorylated version of the normal tau protein, is found in high abundance in NFTs and is therefore more relevant to the pathological development of the disease. There are specific immunoassays for the different phosphorylated epitopes of tau, including threonine 181 (p-tau_{181}) [60], serine 199 (p-tau_{199}) [61], threonine 231 (p-tau_{231}) [62], threonine 231 and

Table 1. Relation between baseline risk factors and risk of conversion to AD in MCI patients; pathological CSF at baseline was associated with the greatest risk. Reproduced with permission from [54]

	Unadjusted hazard ratio (95% Cl)	Adjusted hazard ratio (95% Cl)*
Pathological CSF (T-tau and Aβ42)	30·0 (9·32-96·8)†	17·7 (5·33-58·9)†
Pathological CSF (P-tau$_{181}$ and Aβ42)	26·3 (8·16-84·5)†	16·8 (5·02-56·5)†
Pathological CSF (T-tau and Aβ42/P-tau$_{181}$)	32·8 (10·2-105·6)†	19·8 (5·99-65·7)†
Age, years	1·11 (1·07-1·16)†	1·10 (1·06-1·15)†
Female sex	2·37 (1·33-4·22)‡	1·90 (1·06-3·41)§
APOE ε4 carrier	2·76 (1·50-5·07)‡	2·61 (1·41-4·80)°
Higher education	0·65 (0·37-1·13)	0·80 (0·45-1·43)
Systolic BP, mmHg	1·00 (0·99-1·01)	1·00 (0·98-1·01)
Diastolic BP, mmHg	0·97 (0·95-0·99)ǁ	0·99 (0·97-1·02)
Plasma homocysteine, μM	1·03 (1·04-1·13)°	1·08 (1·04-1·12)**
MMSE, total score	0·87 (0·75-1·02)	1·02 (0·85-1·21)
MMSE, delayed recall	0·67 (0·51-0·88)††	0·85 (0·64-1·13)
Clock-drawing test	0·98 (0·88-1·09)	0·94 (0·84-1·04)

AD = Alsheimer's disease; BP = blood pressure; MMSE = mini-mental state examination. † p,0·0001; ‡ p = 0·001; § p = 0·013; ° p = 0·002; ǁ p = 0·035; ** p = 0005; †† p = 0·009. All data were collected at baseline (t = 0). Cut-off values for pathological CSF were: T-tau > 350 ng/l; Aβ42 < 530 ng/L; P-tau$_{181}$ > 60 ng/L, A β42/P-tau$_{181}$ ratio < 6·5.
* Adjusted (if applicable) for the baseline demographic variables age, sex, education level, and APOE ε4 carrier status

serine 235 (p-tau$_{231\text{-}235}$) [61], and serine 396 and serine 404 (p-tau$_{396/404}$) [63]. P-tau$_{231}$ has been shown to discriminate between AD and other neurological disorders with an overall accuracy of 91% (sensitivity 85%, specificity 97%) [62], and in a separate study when AD was compared to major depression, p-tau$_{231}$ was able to correctly allocate 87% of all cases [64]. Sensitivity levels between AD and FTD improved from 57.7% to 90.2% with p-tau$_{231}$ compared with t-tau, while the specificity for both markers was 92.3% [65]. As a predictive marker, elevated levels of p-tau$_{231}$ at baseline correlated significantly with future cognitive decline (i.e., rate of change in Mini Mental State Examination (MMSE) score) and declined with progression in AD [66]. Likewise, the other phospho-epitope biomarkers, p-tau$_{181, 199, 396/404}$, have also been shown to improve the diagnostic accuracy between AD and other non-AD dementias (VaD, LBD, FTD, and CJD), along with other neurological disorders, such as stroke, CJD, PD, and ALS [61, 63, 67, 68]. Furthermore, p-tau$_{231\text{-}235}$ has been shown to identify the progression of MCI into AD from MCI in 65% of cases and to exclude 100% cases with only memory complaints [69]. Also, restricted analyses of the MCI group showed that reductions in MRI hippocampal volume measures were closely associated with increasing p-tau$_{231}$ levels, suggesting a hyperphosphorylated tau-related progression in AD pathology [70] (Fig. 6).

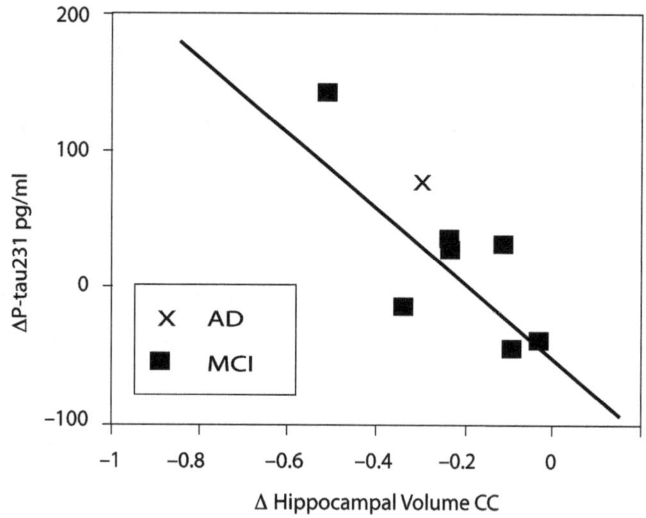

Fig. 6. Correlation between 1-year changes in hippocampal volumes and the changes in p-tau$_{231}$. Reproduced with permission from [6]

In summary, t-tau and p-tau (more closely related to AD pathology) are good diagnostic and prognostic markers in AD. There is also increasing evidence to support a role for tau in diagnosing incipient AD in MCI patients, with relatively high accuracy.

14-3-3 Proteins

14-3-3 isoforms (named β, ε, ζ, η, τ, σ, and θ) are abundantly expressed in the mammalian brain and highly conserved among eukaryotes. Like t-tau, 14-3-3 is released during acute neuronal injury. CJD, viral encephalitis, acute infarction, MS, cerebral neoplasias, and paraneoplastic syndromes have detectable 14-3-3, and occasionally AD, FTD, LBD, and vascular dementia have also been found to be positive [71, 72]. The highest specificity and sensitivity has been reported in sporadic CJD (sCJD) patients, greater than 92% and 96%, respectively [73, 74], whereas including other clinically probable cases in the differential reduces the overall specificity of the assay to 88%, primarily due to conditions such as stroke and viral encephalitis, which can temporarily clinically mimic CJD [75]. False negative results have also been reported early and late in the course of the disease [74]. The WHO has revised its initial diagnostic criteria for sCJD to include a positive 14-3-3 value in patients with progressive dementia of less than 2 years duration [76] (Fig. 7). Therefore, in the correct clinical context, in vivo detection of 14-3-3 and/or a characteristic EEG recording (periodic sharp wave complexes; PSWCs) is considered diagnostic of sCJD.

In the familial forms of spongiform encephalopathies (Gerstmann-Sträusler (P102L) and fatal familial insomnia (D178-129M)), 14-3-3 is largely absent, per-

Progressive dementia;

and

At least 2 out of the following 4 clinical features:

- Myoclonus
- Visual or cerebellar disturbance
- Pyramidal/extrapyramidal dysfunction
- Akinetic mutisni;

and

- A typical EEG during an illness of any duration

and/or

- A positive 14-3-3 CSF assay and a clinical duration to death < 2 years;
- Routine investigations should not suggest an alternative diagnosis.

Fig. 7. Criteria for probable CJD as proposed by the World Health Organization. Reproduced with permission from [76]

haps correlating with the specific mutations [77]. For instance, in familial cases bearing codon 200 (E200K) or 210 (V210I), detection of 14-3-3 is of similar sensitivity and specificity as in sCJD. Equally, in variant CJD (vCJD), a positive 14-3-3 is only found in approximately 86% of patients [78]. This may be improved upon by combining with CSF tau (positive predictive value (PPV) 91% and negative predictive value (NPV) 84%), whereas the combination of 14-3-3 and $A\beta_{42}$ has been found to result in the highest diagnostic accuracy – sensitivity 100%, specificity 98%, PPV 93%, and NPV 100% [74]. Measurement of 14-3-3 is relatively more accurate than other paraclinical measures, for instance EEG has a sensitivity of 66% for sCJD and PSWCs are never seen in vCJD, and the typical heightened basal ganglia activity on MRI is seen in only 78% of vCJD cases [79].

Overall, 14-3-3 is a good diagnostic marker for sCJD and is already widely used clinically in the diagnosis of suspected cases. However, it is not sufficiently specific or sensitive to be used as a diagnostic test in familial CJD or vCJD, and both would still need to be verified pathologically.

Biological Markers of Oxidative Stress Injury

Free radical oxidative stress and mitochondrial dysfunction have been implicated in the final demise of neurons in the majority of neurodegenerative disorders. Nowhere is this more relevant or contentious than in ALS/motor neuron disease (MND). Evidence from transgenic mouse models expressing SOD1 locus (which encodes Cu/Zn-superoxide dismutase) mutations suggest that the altered Cu/Zn-

SOD enzyme activity predisposes motor neurons to oxidative stress injury largely through mishandling of free radicals. This has damaging consequences on cellular proteins, lipids, and DNA, and the animals show signs of slow progressive motor neuron death that closely mimics the human inherited form of the disease. Various analytical methods have been developed to detect oxidative injury; but the more stable forms are modified cellular proteins arising from the oxidative injury.

3-Nitrotyrosine

A prominent mediator of oxidative stress in biological systems is peroxynitrite, which is formed by the reaction of nitric oxide with superoxide. 3-Nitrotyrosine (3-NT) is a marker of peroxynitrite-mediated nitration of tyrosine residues on proteins. Increased concentrations of 3-NT have been reported in the spinal cord of both sporadic and familial patients with ALS, as well as transgenic mouse models [80, 81]. In a CSF study of patients with sporadic ALS (SALS), 3-NT levels and 3-NT/tyrosine ratios were approximately seven times higher in the SALS patients compared with controls. However, no significant correlation was found with either disease duration or severity [82]. In a separate study, the same group demonstrated a significant elevation in free 3-NT levels in AD patients compared with control individuals [83]. Increased 3-NT immunostaining has also been noted in patients with cerebral ischemia, albeit in a more diffuse topographic distribution than that of patients with ALS, where the staining is largely restricted to the ventral horn neurons [80] (Fig. 8). More recently, however, CSF 3-NT levels were found to be elevated in only a few patients with AD or ALS, and the majority of the patients were found to have levels in the same range as the control individuals [84]. This study highlights the importance of case selection and standardization of protocols in biomarker studies.

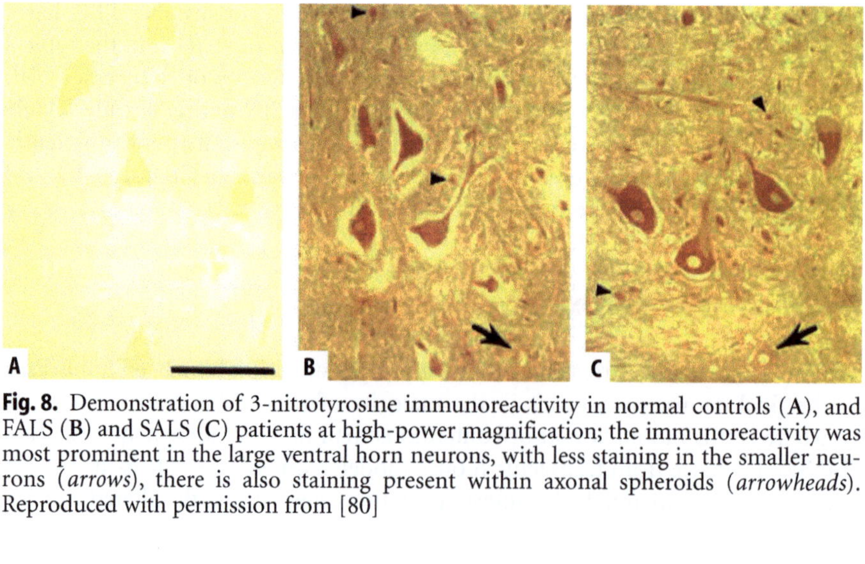

Fig. 8. Demonstration of 3-nitrotyrosine immunoreactivity in normal controls (**A**), and FALS (**B**) and SALS (**C**) patients at high-power magnification; the immunoreactivity was most prominent in the large ventral horn neurons, with less staining in the smaller neurons (*arrows*), there is also staining present within axonal spheroids (*arrowheads*). Reproduced with permission from [80]

Protein Carbonyl Groups

Protein carbonyl groups are formed by oxidative modification of lysine, proline, arginine, and theronine residues on the side-chains of proteins. In humans, the carbonyl content of proteins increases with age, but certain proteins appear more susceptible to protein carbonyl formation than others, such as glutamine synthase and creatinine kinase BB in the brain [85, 86]. Neurodegenerative disorders per se, however, appear to have an excess of modified carbonyl proteins [87, 88]. Disorders in which this has been described include ALS, AD, Pick's disease, LBD, and PD. In SALS, one study documented a rise of 119% in mean protein carbonyl levels in the spinal cord compared with normal control subjects and an 88% rise compared with neurological control subjects [89] (Fig. 9); whilst in the frontal cortex, two other studies documented an 84.8% [90] and 47% [91] rise in protein carbonyl content compared with normal control subjects and disease control subjects, respectively. In subjects with AD, the protein carbonyl content was increased in both the hippocampus and the inferior parietal lobule, corresponding to regions abundant in senile plaques and neurofibrillary tangles [92]. However, in PD, carbonyl levels were found to be increased, although unexpectedly in the cerebellum and the frontal cortex as well as the substantia nigra, caudate nucleus, putamen, and globus pallidus – areas of the brain which are conventionally thought to be involved in PD, suggesting a more widespread oxidative insult to the brain than previously anticipated [93].

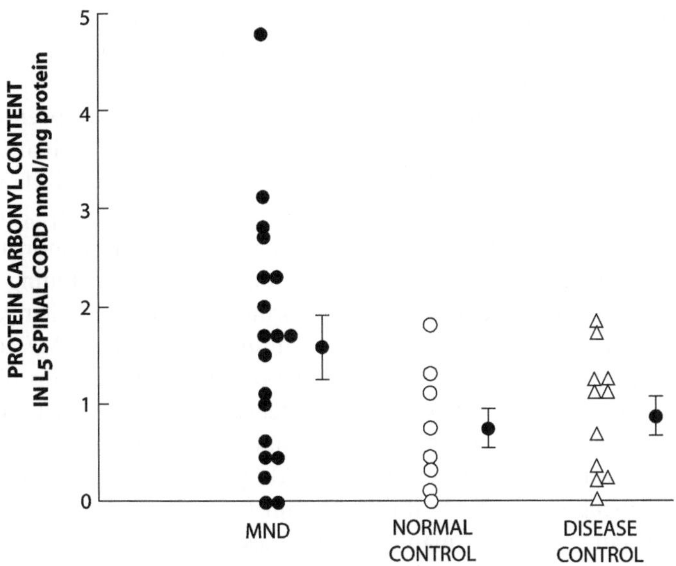

Fig. 9. The protein carbonyl content in the L5 spinal cord samples in the MND group, normal control group, and neurological disease control group. Reproduced with permission from [89]

In summary, although the initial work on oxidative injury markers appears promising, more work is needed in this area in terms of establishing reproducible parameters and in defining the clinical importance of the findings.

Biological Markers of Apoptosis

Cell death is the concluding event in many neurodegenerative disorders, and evidence of neuronal apoptosis has been reported in ALS, PD, AD, HD, CJD, and multiple system atrophy (MSA) [94-100]. The key molecular players in the cell death cascade are the caspases, Fas cell surface ligand receptor system, and the Bcl-2 apoptosis modulating proteins.

Caspase Family

The caspases are a unique group of cysteine proteases that are activated in most cases of apoptotic degradation. Functionally they are divided into initiator caspases at the upstream level (caspases 2, 8, 9, and 10) and executioner caspases at the downstream level, which eventually result in the execution of cell death (caspases 3, 6, and 7).

In ALS, Martin [101] reported increased caspase-3 activity in spinal cord anterior horn and motor cortex, but not in the somatosensory cortex, compared with controls. Ilzecka et al. [102] demonstrated a significant elevation in soluble caspase-1 levels in the serum of ALS patients, with a significant decrease in the CSF compared with controls. The interpretation of the CSF findings from this study is limited by the fact that the control group consisted of patients with lumbosacral disc disease [102].

In AD, cleavage of amyloid precursor protein by caspase has been linked to increased Aβ production, in which caspase-3 is the predominant caspase involved in APP cleavage [103]. Caspase-3-deficient mice show little neuronal loss in the hippocampus following in vivo microinjection of $A\beta_{1-40}$ compared with wild-type mice [104], and in AD patients, caspase-3 immunoreactivity is elevated in dying pyramidal neurons of the hippocampus, including the CA3 region, which is one of the sites of initial neuronal loss [103]. Similarly, Masliah et al. [105] showed increased caspase-3 immunoreactivity in AD compared with control brain, and the staining was more intense in neurons displaying DNA fragmentation [105]. More recent evidence also suggests that tau may be targeted by caspases [106, 107], which may play an important role in the development of the neurofibrillary tangles as AD progresses. In support of this hypothesis, using an antibody directed against the caspase-cleaved carboxyl-terminus of tau (Δtau), Rissman et al. [108] also demonstrated that Δtau was present in brain extracts of patients with MCI and AD but not controls, suggesting that Δtau production may coincide with early stages of cognitive decline in AD (Fig. 10).

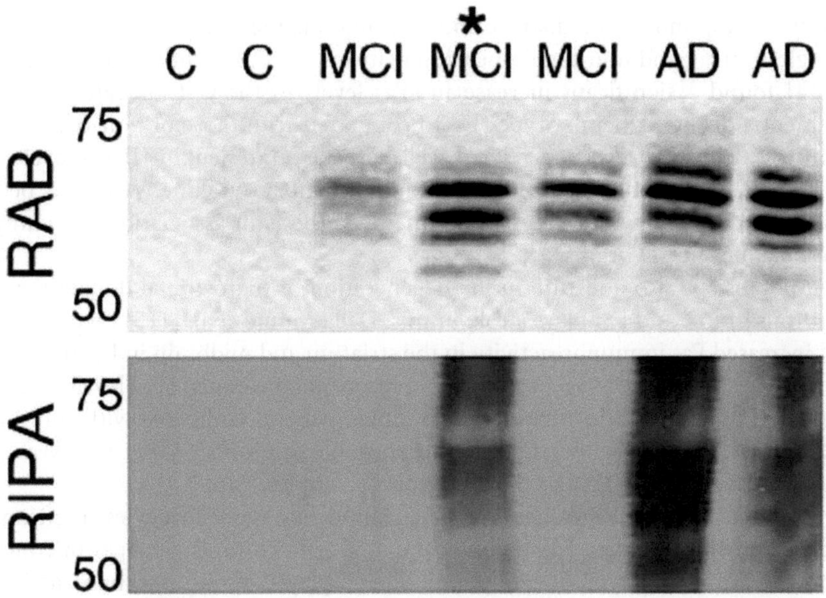

Fig. 10. Caspase-cleaved tau (ΔTau) becomes increasingly insoluble with AD disease progression. ΔTau was detected in the soluble high-salt RAB fraction of MCI and AD cases but not controls (*C*), suggesting that ΔTau production coincides with the early stages of cognitive decline in AD. In detergent-soluble RIPA fractions, ΔTau was only observed in higher pathology MCI and AD cases, suggesting that ΔTau becomes more insoluble with AD progression. The *multiple bands* correspond to tau isoforms. *This patient was transitional between MCI and AD. Reproduced with permission from [108]

Aside from AD, caspase-cleaved fragments can also be found in HD in the form of expanded mutant huntingtin (htt) proteins that are detectable in the brain tissue of affected HD patients, as well as in early grade HD postmortem brains [109, 110]. More recently, Graham et al. [111] demonstrated that the caspase-6 cleavage site in mutant htt was the rate-limiting event in the pathogenesis of HD, and selective elimination of the cleavage site provides sufficient protection from neurotoxicity induced by a variety of stressors including NMDA, and prevents subsequent striatal neurodegeneration in HD [111].

Fas

Fas receptor ligand (FasL) and soluble Fas (sFas) belong to the nerve growth factor-tumor necrosis factor-α receptor family (NGF/TNF-α) of apoptosis-signaling molecules.

In AD, Fas-positive astrocytes occur almost exclusively in affected areas of the brain, most frequently the hippocampus [112]. Areas of Fas immunoreactivity correspond to the distribution of senile plaques and are found in both tangle-bearing and non-tangle-bearing neurons [113]. Recently, Martinez et al. [114] found a significant increase in sFas levels in the CSF of patients with dementia of the Alzheimer type (DAT) compared with those of non-demented controls. The sFas levels correlated well with CSF interleukin (IL)-6 levels in DAT patients ($r = 0.703$; $P < 0.05$), an inflammatory cytokine which has been implicated in β-amyloid precursor protein synthesis in association with amyloid plaques in AD brains.

Apoptosis is also a feature in the degeneration of nigrostriatal dopaminergic neurons in PD, and Fas plays a role in this. De la Monte et al. [115] demonstrated increased Fas immunoreactivity in the striatum and midbrain in PD histologic sections. Similarly, Mogi et al. [116] demonstrated elevated sFas levels in caudate and putamen brain areas of parkinsonian patients compared with controls. However, the investigators found no detectable amounts (<16 pg/L) sFas in either the ventricular CSF or the lumbar CSF samples [116].

Elevated Fas expression has also been found in oligodendrocytes in chronic active and chronic silent MS lesions compared with controls, whereas microglia and the infiltrating lymphocyte population displayed strong immunoreactivity to FasL. Fas coupling with FasL induced rapid oligodendrocyte membrane lysis [117]. Similarly, Bilinska et al. [118] found greater activation of Fas on the surface of CD4+ lymphocyte subset during a relapse than during remission. Elevated sFas levels have also been found in the sera and CSF of patients with active MS [119], and longitudinally high sFas levels were related to EDSS changes and hence disease activity [120].

Bcl-2 Family

The Bcl-2 family comprises pro-apoptotic (e.g., Bax, Bad, Bak, and Bcl-xS) and anti-apoptotic (e.g., Bcl-2 and Bcl-xL) proteins, their relative ratios ultimately dictating the sensitivity of cells to various apoptotic signals.

In ALS, the balance of pro- and anti-apoptotic Bcl-2 family proteins is altered in favor of apoptosis. In both symptomatic mutant transgenic SOD1 mice and human ALS cases, the expression of anti-apoptotic proteins Bcl-2 and Bcl-X_L was found to be either unchanged or decreased in the spinal cord motor neurons, whereas pro-apoptotic proteins Bax and Bad were found to be increased [101, 121]. A similar shift toward apoptosis has also been observed in AD and PD. In the AD brain, there is downregulation of Bcl-2 in tangle-bearing neurons, with prominence of Bax-containing neurons in the hippocampus [122-124], whereas in PD there is increased Bax immunoreactivity in nigral neurons compared with controls [125].

In summary, as apoptosis is the penultimate step leading to neuronal death in neurodegeneration, therapies with demonstrative neuroprotective function can be evaluated using apoptotic markers.

Other Biological Markers

B-Amyloid Peptides

B-Amyloid peptides (Aβ) form an integral part of senile amyloid plaques in AD. Aβ forms following enzymatic cleavage of the parent amyloid precursor protein and the Aβ42-isoform has been found to have the greatest diagnostic potential for the accurate diagnosis of AD. A consistent finding from most of the studies is a decrease in the level of Aβ42 to approximately 50% of that found in AD controls [126]; a finding which is thought to result at least partly from the deposition of Aβ within plaques in the brain, with only some of it finding its way into the CSF.

The mean sensitivity of CSF-Aβ42 is approximately 86%, while the specificity is around 91% in discriminating AD from normal aging [126]. Like tau, Aβ peptides are unable to differentiate AD clearly from other dementias (LBD, VAD, frontotemporal dementia), and likewise other neurological disorders (CJD, ALS, acute bacterial meningitis) [48, 127-130]. However, combining the CSF Aβ-isoforms Aβ42/Aβ40 (or vice versa) as a ratio has shown greater discriminative power than CSF-Aβ42 alone in making the diagnosis of AD. Unlike Aβ42, Aβ40 remains unchanged in the CSF, and as a consequence the reduction in Aβ42/Aβ40 ratio is more marked than the reduction in Aβ42 alone. Performance-wise, the Aβ42/Aβ40 ratio correctly classified more patients with AD than Aβ42 alone: AD vs non-AD dementia; 90% and 85%; and AD vs non-AD dementia plus controls; 90.8% and 87%, respectively [131] (Fig. 11). In longitudinal evaluation, the Aβ42/Aβ40 ratio increases over the follow-up period, improving the sensitivity of the test [57].

Neuron-Specific Enolase

Neuron-specific enolase (NSE) is a dimeric glycolytic enzyme of neuronal cells (γγ dimer) and astrocytes (αα dimer), synthesized almost exclusively in the CNS. CSF NSE levels rise following rapid neuronal destruction, and elevated levels have been documented in several neurodegenerative disorders, including CJD, AD, VAD, and MS [132-135]. It has a potential use as a non-disease-specific marker of neurodegeneration.

Glial Fibrillary Acidic Protein

Glial fibrillary acidic protein (GFAP) is a marker of astroglial activation in response to neuronal injury. Astroglial activation is often rapid, and marked increases in GFAP have been observed in acute neurological disorders such as stroke and traumatic brain injury [136-139]. Studies in more chronic neurological disorders have found elevated GFAP concentrations in patients with dementia (AD and VAD) [140, 141], syringomyelia [136], and hydrocephalus [142]. Elevated levels of GFAP have also been reported in the progressive phase of MS, suggesting that GFAP may be marker for irreversible damage in MS [24, 143, 144].

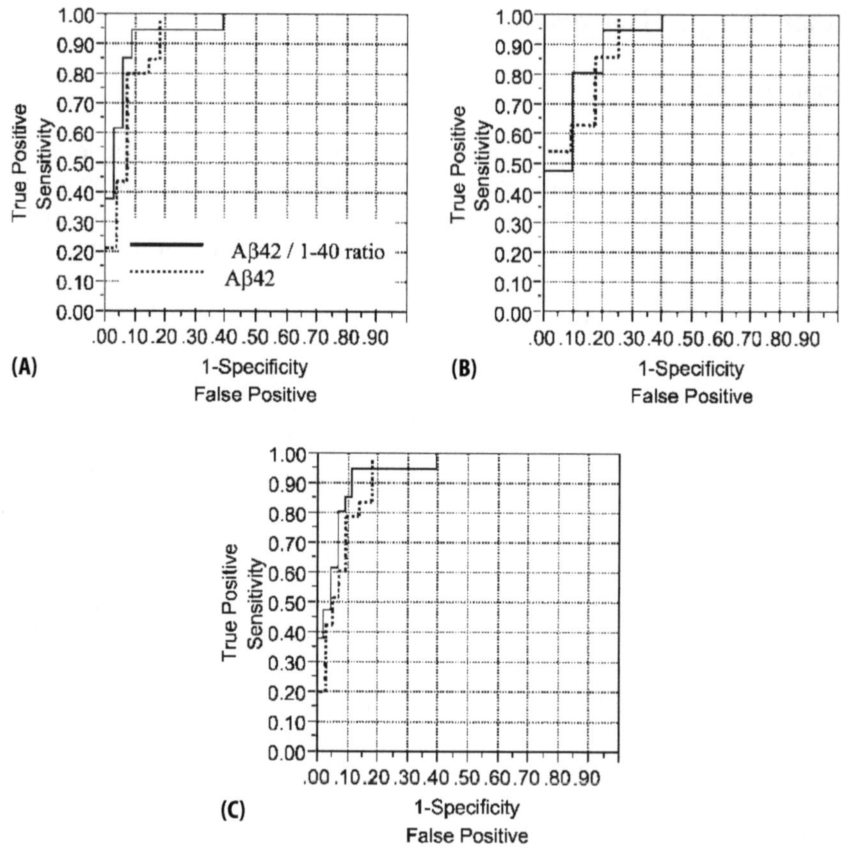

Fig. 11. Receiver operating curves (ROC) for Aβ42 and Aβ peptides ratio for discriminating AD from a control group (**A**), non-AD dementia group (**B**), and both groups taken together (**C**). Reproduced with permission from [131]

Limitations to the Use of Biological Markers in Everyday Clinical Practice

The search for more sensitive and specific biomarkers for use in the diagnosis and assessment of neurodegenerative disorders should be tempered by an understanding of the limitations of the technique and methodology, the sample quality, and the selection of controls. Due to differences in all of the above, there are many conflicting reports on biological markers in the literature. The use of standardized protocols and reporting guidelines would help to reduce the disparity. Oncologists have looked at this in great detail and a copy of their guidelines can be found in peer-reviewed journals [145].

Aside from the technical limitations, the more pertinent limitations to the use of biomarkers in everyday clinical practice are as follows:

1. Performance of biomarkers outside the research environment and, to a lesser extent, outside specialist units;
2. Reliance on the accuracy of clinical diagnosis and evaluation before performing the test. Visser et al. [146] showed that the diagnostic accuracy of MCI criteria used in previous drug trials to identify patients with pre-dementia was low to moderate: the sensitivity varied between 0.46 and 0.83, whilst the PPV ranged between 0.43 and 0.76 [146].
3. The necessity for a lumbar puncture, particularly in those with impaired capacity to consent. However, in a study of 235 patients undergoing research lumbar punctures, the incidence of an adverse event was 11.7%, a clinically significant adverse event 3.97%, and a typical post-LP headache only 0.93%;
4. The potential discovery of a single useful biological marker is becoming increasingly elusive, as most neurodegenerative disorders share common pathogenic mechanisms. A selective biomarker panel algorithm may therefore be more predictive of a disorder.

As molecular biomarkers become more widely adopted, the above limitations should become less of a problem in everyday clinical practice.

Conclusions

Within the last decade or so there has been an explosion in the number of publications on biological markers. The concept has been reworked, reintroduced, and discarded on numerous occasions, but whatever the flaws, it is still more promising than any other approach we currently have in practice. For neurodegenerative disorders, biological markers can provide useful information on the stage and state of the underlying disease pathogenesis more accurately and at an earlier stage, ultimately providing more precision clinically and the ability to detect incipient disease in patients.

In this chapter, we have provided an overview of biological markers and their application in the diagnosis and prognosis of selected neurodegenerative conditions. They have been categorized into those of axonal injury and acute neurodestruction, oxidative stress injury, and apoptosis. Many of the biological markers have achieved a status of high sensitivity with reasonable specificity, but no single biomarker has been found to be specific for a particular condition. Unless the pathogenic step is unique to that particular disorder, the chances of finding a single biological marker for that disorder are slim. In such circumstances, a biomarker panel algorithm may be more useful. Taking AD as an example, CSF $A\beta_{42}$ and tau could be used as diagnostic biomarkers, whilst Nf and the caspases could be the prognostic biomarkers. Moreover, combining biological markers with other paraclinical assessments, such as imaging and neurophysiology, should provide even greater sensitivity and specificity.

Acknowledgements. We thank the National MS Society of the United States and the MS Society of Great Britain and Northern Ireland for their support, and Ms. Janet Alsop and Dr. Geoff Keir for their critical review of the chapter.

References

1. Ferguson B, Matyszak MK, Esiri MM, Perry VH (1997) Axonal damage in acute multiple sclerosis lesions. Brain 120(3):393-399
2. Carden MJ, Schlaepfer WW, Lee VM (1985) The structure, biochemical properties, and immunogenicity of neurofilament peripheral regions are determined by phosphorylation state. J Biol Chem 260:9805-9817
3. Perrone Capano C, Pernas-Alonso R, di Porzio U (2001) Neurofilament homeostasis and motoneurone degeneration. Bioessays 23:24-33
4. Petzold A (2005) Neurofilament phosphoforms: surrogate markers for axonal injury, degeneration and loss. J Neurol Sci 233:183-198
5. Carpenter S (1968) Proximal axonal enlargement in motor neuron disease. Neurology 18:841-851
6. Hirano A, Donnenfeld H, Sasaki S, Nakano I (1984) Fine structural observations of neurofilamentous changes in amyotrophic lateral sclerosis. J Neuropathol Exp Neurol 43:461-470
7. Hirano A, Nakano I, Kurland LT et al (1984) Fine structural study of neurofibrillary changes in a family with amyotrophic lateral sclerosis. J Neuropathol Exp Neurol 43:471-480
8. Dickson DW, Feany MB, Yen SH et al (1996) Cytoskeletal pathology in non-Alzheimer degenerative dementia: new lesions in diffuse Lewy body disease, Pick's disease, and corticobasal degeneration. J Neural Transm 47(Suppl):31-46
9. de la Monte SM, Wands JR (1994) Diagnostic utility of quantitating neurofilament-immunoreactive Alzheimer's disease lesions. J Histochem Cytochem 42:1625-1634
10. Arima K, Nakamura M, Sunohara N et al (1999) Immunohistochemical and ultrastructural characterization of neuritic clusters around ghost tangles in the hippocampal formation in progressive supranuclear palsy brains. Acta Neuropathol (Berl) 97:565-576
11. Brettschneider J, Petzold A, Sussmuth SD et al (2006) Axonal damage markers in cerebrospinal fluid are increased in ALS. Neurology 66:852-856
12. Norgren N, Rosengren L, Stigbrand T (2003) Elevated neurofilament levels in neurological diseases. Brain Res 987:25-31
13. Rosengren LE, Karlsson JE, Karlsson JO et al (1996) Patients with amyotrophic lateral sclerosis and other neurodegenerative diseases have increased levels of neurofilament protein in CSF. J Neurochem 67:2013-2018
14. Sjogren M, Blomberg M, Jonsson M et al (2001) Neurofilament protein in cerebrospinal fluid: a marker of white matter changes. J Neurosci Res 66:510-516
15. Lovas G, Szilagyi N, Majtenyi K et al (2000) Axonal changes in chronic demyelinated cervical spinal cord plaques. Brain 123(2):308-317
16. Trapp BD, Peterson J, Ransohoff RM, Rudick R, Mork S, Bo L (1998) Axonal transection in the lesions of multiple sclerosis. N Engl J Med 338:278-285
17. Norgren N, Sundstrom P, Svenningsson A et al (2004) Neurofilament and glial fibrillary acidic protein in multiple sclerosis. Neurology 63:1586-1590
18. Petzold A, Eikelenboom MJ, Keir G et al (2005) Axonal damage accumulates in the progressive phase of multiple sclerosis: three year follow up study. J Neurol Neurosurg Psychiat 76:206-211

19. Petzold A, Eikelenboom MI, Keir G et al (2006) The new global multiple sclerosis severity score (MSSS) correlates with axonal but not glial biomarkers. Mult Scler 12:325-328
20. Petzold A, Rejdak K, Plant GT (2004) Axonal degeneration and inflammation in acute optic neuritis. J Neurol Neurosurg Psychiat 75:1178-1180
21. Brettschneider J, Petzold A, Junker A, Tumani H (2006) Axonal damage markers in the cerebrospinal fluid of patients with clinically isolated syndrome improve predicting conversion to definite multiple sclerosis. Mult Scler 12:143-148
22. Griffin WS, Sheng JG, McKenzie JE et al (1998) Life-long overexpression of S100beta in Down's syndrome: implications for Alzheimer pathogenesis. Neurobiol Aging 19:401-405
23. Otto M, Wiltfang J, Schutz E et al (1998) Diagnosis of Creutzfeldt-Jakob disease by measurement of S100 protein in serum: prospective case-control study. BMJ 316:577-582
24. Petzold A, Eikelenboom MJ, Gveric D et al (2002) Markers for different glial cell responses in multiple sclerosis: clinical and pathological correlations. Brain 125:1462-1473
25. Foerch C, Otto B, Singer OC et al (2004) Serum S100B predicts a malignant course of infarction in patients with acute middle cerebral artery occlusion. Stroke 35:2160-2164
26. Infante JR, Martinez A, Ochoa J et al (2003) Cerebrospinal fluid S-100 protein levels in neurological pathologies. J Physiol Biochem 59:255-261
27. Petzold A, Jenkins R, Watt HC et al (2003) Cerebrospinal fluid S100B correlates with brain atrophy in Alzheimer's disease. Neurosci Lett 336:167-170
28. Peskind ER, Griffin WS, Akama KT et al (2001) Cerebrospinal fluid S100B is elevated in the earlier stages of Alzheimer's disease. Neurochem Int 39:409-413
29. Green AJ, Harvey RJ, Thompson EJ, Rossor MN (1997) Increased S100beta in the cerebrospinal fluid of patients with frontotemporal dementia. Neurosci Lett 235:5-8
30. Otto M, Bahn E, Wiltfang J et al (1998) Decrease of S100 beta protein in serum of patients with amyotrophic lateral sclerosis. Neurosci Lett 240:171-173
31. Sussmuth SD, Tumani H, Ecker D, Ludolph AC (2003) Amyotrophic lateral sclerosis: disease stage related changes of tau protein and S100 beta in cerebrospinal fluid and creatine kinase in serum. Neurosci Lett 353:57-60
32. Michetti F, Massaro A, Murazio M (1979) The nervous system-specific S-100 antigen in cerebrospinal fluid of multiple sclerosis patients. Neurosci Lett 11:171-175
33. Massaro AR, Michetti F, Laudisio A, Bergonzi P (1985) Myelin basic protein and S-100 antigen in cerebrospinal fluid of patients with multiple sclerosis in the acute phase. Ital J Neurol Sci 6:53-56
34. Missler U, Wandinger KP, Wiesmann M et al (1997) Acute exacerbation of multiple sclerosis increases plasma levels of S-100 protein. Acta Neurol Scand 96:142-144
35. Lim ET, Petzold A, Leary SM et al (2004) Serum S100B in primary progressive multiple sclerosis patients treated with interferon-beta-1a. J Negat Results Biomed 3:4
36. Buttner T, Weyers S, Postert T, Sprengelmeyer R, Kuhn W (1997) S-100 protein: serum marker of focal brain damage after ischemic territorial MCA infarction. Stroke 28:1961-1965
37. Aurell A, Rosengren LE, Karlsson B et al (1991) Determination of S-100 and glial fibrillary acidic protein concentrations in cerebrospinal fluid after brain infarction. Stroke 22:1254-1258
38. Missler U, Wiesmann M, Friedrich C, Kaps M (1997) S-100 protein and neuron-specific enolase concentrations in blood as indicators of infarction volume and prognosis in acute ischemic stroke. Stroke 28:1956-1960

39. Blennow K, Vanmechelen E (2003) CSF markers for pathogenic processes in Alzheimer's disease: diagnostic implications and use in clinical neurochemistry. Brain Res Bull 61:235-242
40. Blennow K, Hampel H (2003) CSF markers for incipient Alzheimer's disease. Lancet Neurol 2:605-613
41. Hampel H, Goernitz A, Buerger K (2003) Advances in the development of biomarkers for Alzheimer's disease: from CSF total tau and Abeta(1-42) proteins to phosphorylated tau protein. Brain Res Bull 61:243-253
42. Hampel H, Mitchell A, Blennow K et al (2004) Core biological marker candidates of Alzheimer's disease: perspectives for diagnosis, prediction of outcome and reflection of biological activity. J Neural Transm 111:247-272
43. Morikawa Y, Arai H, Matsushita S et al (1999) Cerebrospinal fluid tau protein levels in demented and nondemented alcoholics. Alcohol Clin Exp Res 23:575-577
44. Andreasen N, Minthon L, Clarberg A et al (1999) Sensitivity, specificity, and stability of CSF-tau in AD in a community-based patient sample. Neurology 53:1488-1494
45. Molina JA, Benito-León J, Jiménez-Jiménez FJ et al (1997) Tau protein concentrations in cerebrospinal fluid of non-demented Parkinson's disease patients. Neurosci Lett 238:139-141
46. Hulstaert F, Blennow K, Ivanoiu A et al (1999) Improved discrimination of AD patients using beta-amyloid(1-42) and tau levels in CSF. Neurology 52:1555-1562
47. Paraskevas GP, Kapaki E, Liappas I et al (2005) The diagnostic value of cerebrospinal fluid tau protein in dementing and nondementing neuropsychiatric disorders. J Geriatr Psychiatry Neurol 18:163-173
48. Andreasen N, Minthon L, Davidsson P et al (2001) Evaluation of CSF-tau and CSF-Abeta42 as diagnostic markers for Alzheimer disease in clinical practice. Arch Neurol 58:373-379
49. Otto M, Wiltfang J, Tumani H et al (1997) Elevated levels of tau-protein in cerebrospinal fluid of patients with Creutzfeldt-Jakob disease. Neurosci Lett 225:210-212
50. Hesse C, Rosengren L, Vanmechelen E et al (2000) Cerebrospinal fluid markers for Alzheimer's disease evaluated after acute ischemic stroke. J Alzheimers Dis 2:199-206
51. Brettschneider J, Maier M, Arda S et al (2005) Tau protein level in cerebrospinal fluid is increased in patients with early multiple sclerosis. Mult Scler 11:261-265
52. Bulut M, Koksal O, Dogan S et al (2006) Tau protein as a serum marker of brain damage in mild traumatic brain injury: preliminary results. Adv Ther 23:12-22
53. Arai H, Nakagawa T, Kosaka Y et al (1997) Elevated cerebrospinal fluid tau protein level as a predictor of dementia in memory-impaired patients. Alzheim Res 3:211-213
54. Hansson O, Zetterberg H, Buchhave P et al (2006) Association between CSF biomarkers and incipient Alzheimer's disease in patients with mild cognitive impairment: a follow-up study. Lancet Neurol 5:228-234
55. Maruyama M, Arai H, Sugita M et al (2001) Cerebrospinal fluid amyloid beta(1-42) levels in the mild cognitive impairment stage of Alzheimer's disease. Exp Neurol 172:433-436
56. Hock C, Golombowski S, Naser W, Muller-Spahn F (1995) Increased levels of tau protein in cerebrospinal fluid of patients with Alzheimer's disease: correlation with degree of cognitive impairment. Ann Neurol 37:414-415
57. Kanai M, Matsubara E, Isoe K et al (1998) Longitudinal study of cerebrospinal fluid levels of tau, A beta1-40, and A beta1-42(43) in Alzheimer's disease: a study in Japan. Ann Neurol 44:17-26
58. Andreasen N, Vanmechelen E, Van de Voorde A et al (1998) Cerebrospinal fluid tau protein as a biochemical marker for Alzheimer's disease: a community based follow up study. J Neurol Neurosurg Psychiatry 64:298-305

59. Arai H, Terajima M, Miura M et al (1995) Tau in cerebrospinal fluid: a potential diagnostic marker in Alzheimer's disease. Ann Neurol 38:649-652
60. Vanmechelen E, Vanderstichele H, Davidsson P et al (2000) Quantification of tau phosphorylated at threonine 181 in human cerebrospinal fluid: a sandwich ELISA with a synthetic phosphopeptide for standardization. Neurosci Lett 285:49-52
61. Ishiguro K, Ohno H, Arai H et al (1999) Phosphorylated tau in human cerebrospinal fluid is a diagnostic marker for Alzheimer's disease. Neurosci Lett 270:91-94
62. Kohnken R, Buerger K, Zinkowski R et al (2000) Detection of tau phosphorylated at threonine 231 in cerebrospinal fluid of Alzheimer's disease patients. Neurosci Lett 287:187-190
63. Hu YY, He SS, Wang X et al (2002) Levels of nonphosphorylated and phosphorylated tau in cerebrospinal fluid of Alzheimer's disease patients: an ultrasensitive bienzyme-substrate-recycle enzyme-linked immunosorbent assay. Am J Pathol 160:1269-1278
64. Buerger K, Zinkowski R, Teipel SJ et al (2003) Differentiation of geriatric major depression from Alzheimer's disease with CSF tau protein phosphorylated at threonine 231. Am J Psychiatry 160:376-379
65. Buerger K, Zinkowski R, Teipel SJ et al (2002) Differential diagnosis of Alzheimer disease with cerebrospinal fluid levels of tau protein phosphorylated at threonine 231. Arch Neurol 59:1267-1272
66. Buerger K, Teipel SJ, Zinkowski R et al (2002) CSF tau protein phosphorylated at threonine 231 correlates with cognitive decline in MCI subjects. Neurology 59:627-629
67. Sjogren M, Davidsson P, Tullberg M et al (2001) Both total and phosphorylated tau are increased in Alzheimer's disease. J Neurol Neurosurg Psychiat 70:624-630
68. Hesse C, Rosengren L, Andreasen N et al (2001) Transient increase in total tau but not phospho-tau in human cerebrospinal fluid after acute stroke. Neurosci Lett 297:187-190
69. Arai H, Ishiguro K, Ohno H et al (2000) CSF phosphorylated tau protein and mild cognitive impairment: a prospective study. Exp Neurol 166:201-203
70. de Leon MJ, DeSanti S, Zinkowski R et al (2006) Longitudinal CSF and MRI biomarkers improve the diagnosis of mild cognitive impairment. Neurobiol Aging 27:394-401
71. Burkhard PR, Sanchez JC, Landis T, Hochstrasser DF (2001) CSF detection of the 14-3-3 protein in unselected patients with dementia. Neurology 56:1528-1533
72. Tschampa HJ, Neumann M, Zerr I et al (2001) Patients with Alzheimer's disease and dementia with Lewy bodies mistaken for Creutzfeldt-Jakob disease. J Neurol Neurosurg Psychiat 71:33-39
73. Collins S, Boyd A, Fletcher A et al (2000) Creutzfeldt-Jakob disease: diagnostic utility of 14-3-3 protein immunodetection in cerebrospinal fluid. J Clin Neurosci 7:203-208
74. Van EB, Quoilin S, Boons J et al (2003) A prospective study of CSF markers in 250 patients with possible Creutzfeldt-Jakob disease. J Neurol Neurosurg Psychiat 74:1210-1214
75. Hsich G, Kenney K, Gibbs CJ et al (1996) The 14-3-3 brain protein in cerebrospinal fluid as a marker for transmissible spongiform encephalopathies. N Engl J Med 335:924-930
76. WHO (1998) Human transmissible spongiform encephalopathies. Weekly Epidemiol Rec 73:361-372
77. Zerr I, Pocchiari M, Collins S et al (2000) Analysis of EEG and CSF 14-3-3 proteins as aids to the diagnosis of Creutzfeldt-Jakob disease. Neurology 55:811-815
78. Green AJ, Thompson EJ, Stewart GE et al (2001) Use of 14-3-3 and other brain-specific proteins in CSF in the diagnosis of variant Creutzfeldt-Jakob disease. J Neurol Neurosurg Psychiat 70:744-748
79. Zeidler M, Sellar RJ, Collie DA et al (2000) The pulvinar sign on magnetic resonance imaging in variant Creutzfeldt-Jakob disease. Lancet 355:1412-1418

80. Beal MF, Ferrante RJ, Browne SE et al (1997) Increased 3-nitrotyrosine in both sporadic and familial amyotrophic lateral sclerosis. Ann Neurol 42:644-654
81. Ferrante RJ, Shinobu LA, Schulz JB et al (1997) Increased 3-nitrotyrosine and oxidative damage in mice with a human copper/zinc superoxide dismutase mutation. Ann Neurol 42:326-334
82. Tohgi H, Abe T, Yamazaki K et al (1999) Remarkable increase in cerebrospinal fluid 3-nitrotyrosine in patients with sporadic amyotrophic lateral sclerosis. Ann Neurol 46:129-131
83. Tohgi H, Abe T, Yamazaki K et al (1999) Alterations of 3-nitrotyrosine concentration in the cerebrospinal fluid during aging and in patients with Alzheimer's disease. Neurosci Lett 269:52-54
84. Ryberg H, Soderling AS, Davidsson P et al (2004) Cerebrospinal fluid levels of free 3-nitrotyrosine are not elevated in the majority of patients with amyotrophic lateral sclerosis or Alzheimer's disease. Neurochem Int 45:57-62
85. Carney JM, Starke-Reed PE, Oliver CN et al (1991) Reversal of age-related increase in brain protein oxidation, decrease in enzyme activity, and loss in temporal and spatial memory by chronic administration of the spin-trapping compound N-tert-butyl-alpha-phenylnitrone. Proc Natl Acad Sci USA 88:3633-3636
86. Smith CD, Carney JM, Tatsumo T et al (1992) Protein oxidation in aging brain. Ann N Y Acad Sci 663:110-119
87. Aksenova MV, Aksenov MY, Payne RM et al (1999) Oxidation of cytosolic proteins and expression of creatine kinase BB in frontal lobe in different neurodegenerative disorders. Dement Geriatr Cogn Disord 10:158-165
88. Aksenov M, Aksenova M, Butterfield DA, Markesbery WR (2000) Oxidative modification of creatine kinase BB in Alzheimer's disease brain. J Neurochem 74:2520-2527
89. Shaw PJ, Ince PG, Falkous G, Mantle D (1995) Oxidative damage to protein in sporadic motor neuron disease spinal cord. Ann Neurol 38:691-695
90. Bowling AC, Schulz JB, Brown RH, Jr, Beal MF (1993) Superoxide dismutase activity, oxidative damage, and mitochondrial energy metabolism in familial and sporadic amyotrophic lateral sclerosis. J Neurochem 61:2322-2325
91. Ferrante RJ, Browne SE, Shinobu LA et al (1997) Evidence of increased oxidative damage in both sporadic and familial amyotrophic lateral sclerosis. J Neurochem 69:2064-2074
92. Hensley K, Hall N, Subramaniam R et al (1995) Brain regional correspondence between Alzheimer's disease histopathology and biomarkers of protein oxidation. J Neurochem 65:2146-2156
93. Alam ZI, Daniel SE, Lees AJ et al (1997) A generalised increase in protein carbonyls in the brain in Parkinson's but not incidental Lewy body disease. J Neurochem 69:1326-1329
94. Przedborski S (2004) Programmed cell death in amyotrophic lateral sclerosis: a mechanism of pathogenic and therapeutic importance. Neurologist 10:1-7
95. Shaw PJ (2005) Molecular and cellular pathways of neurodegeneration in motor neurone disease. J Neurol Neurosurg Psychiat 76:1046-1057
96. Tatton WG, Chalmers-Redman R, Brown D, Tatton N (2003) Apoptosis in Parkinson's disease: signals for neuronal degradation. Ann Neurol 53(Suppl 3):S61-S70
97. Eckert A, Marques CA, Keil U et al (2003) Increased apoptotic cell death in sporadic and genetic Alzheimer's disease. Ann N Y Acad Sci 1010:604-609
98. Dragunow M, Faull RL, Lawlor P et al (1995) In situ evidence for DNA fragmentation in Huntington's disease striatum and Alzheimer's disease temporal lobes. Neuroreport 6:1053-1057
99. Puig B, Ferrer I (2001) Cell death signaling in the cerebellum in Creutzfeldt-Jakob disease. Acta Neuropathol (Berl) 102:207-215

100. Probst-Cousin S, Rickert CH, Schmid KW, Gullotta F (1998) Cell death mechanisms in multiple system atrophy. J Neuropathol Exp Neurol 57:814-821
101. Martin LJ (1999) Neuronal death in amyotrophic lateral sclerosis is apoptosis: possible contribution of a programmed cell death mechanism. J Neuropathol Exp Neurol 58:459-471
102. Ilzecka J, Stelmasiak Z, Dobosz B (2001) Interleukin-1beta converting enzyme/caspase-1 (ICE/caspase-1) and soluble APO-1/Fas/CD 95 receptor in amyotrophic lateral sclerosis patients. Acta Neurol Scand 103:255-258
103. Gervais FG, Xu D, Robertson GS et al (1999) Involvement of caspases in proteolytic cleavage of Alzheimer's amyloid-beta precursor protein and amyloidogenic A beta peptide formation. Cell 97:395-406
104. Takuma H, Tomiyama T, Kuida K, Mori H (2004) Amyloid beta peptide-induced cerebral neuronal loss is mediated by caspase-3 in vivo. J Neuropathol Exp Neurol 63:255-261
105. Masliah E, Mallory M, Alford M et al (1998) Caspase dependent DNA fragmentation might be associated with excitotoxicity in Alzheimer disease. J Neuropathol Exp Neurol 57:1041-1052
106. Gamblin TC, Chen F, Zambrano A et al (2003) Caspase cleavage of tau: linking amyloid and neurofibrillary tangles in Alzheimer's disease. Proc Natl Acad Sci USA 100:10032-10037
107. Chung CW, Song YH, Kim IK et al (2001) Proapoptotic effects of tau cleavage product generated by caspase-3. Neurobiol Dis 8:162-172
108. Rissman RA, Poon WW, Blurton-Jones M et al (2004) Caspase-cleavage of tau is an early event in Alzheimer disease tangle pathology. J Clin Invest 114:121-130
109. Kim YJ, Yi Y, Sapp E et al (2001) Caspase 3-cleaved N-terminal fragments of wild-type and mutant huntingtin are present in normal and Huntington's disease brains, associate with membranes, and undergo calpain-dependent proteolysis. Proc Natl Acad Sci USA 98:12784-12789
110. Wellington CL, Ellerby LM, Gutekunst CA et al (2002) Caspase cleavage of mutant huntingtin precedes neurodegeneration in Huntington's disease. J Neurosci 22:7862-7872
111. Graham RK, Deng Y, Slow EJ et al (2006) Cleavage at the caspase-6 site is required for neuronal dysfunction and degeneration due to mutant huntingtin. Cell 125:1179-1191
112. Smale G, Nichols NR, Brady DR et al (1995) Evidence for apoptotic cell death in Alzheimer's disease. Exp Neurol 133:225-230
113. Ferrer I, Puig B, Krupinsk J et al (2001) Fas and Fas ligand expression in Alzheimer's disease. Acta Neuropathol (Berl) 102:121-131
114. Martinez M, Fernandez-Vivancos E, Frank A, De la Fuente M, Hernanz A (2000) Increased cerebrospinal fluid fas (Apo-1) levels in Alzheimer's disease: relationship with IL-6 concentrations. Brain Res 869:216-219
115. de la Monte SM, Sohn YK, Ganju N, Wands JR (1998) P53- and CD95-associated apoptosis in neurodegenerative diseases. Lab Invest 78:401-411
116. Mogi M, Harada M, Kondo T et al (1996) The soluble form of Fas molecule is elevated in parkinsonian brain tissues. Neurosci Lett 220:195-198
117. D'Souza SD, Bonetti B, Balasingam V et al (1996) Multiple sclerosis: Fas signaling in oligodendrocyte cell death. J Exp Med 184:2361-2370
118. Bilinska M, Frydecka I, Podemski R, Gruszka E (2003) Fas expression on T cells and sFas in relapsing-remitting multiple sclerosis. Acta Neurol Scand 107:387-393
119. Boylan MT, Crockard AD, McDonnell GV et al (2001) Serum and cerebrospinal fluid soluble Fas levels in clinical subgroups of multiple sclerosis. Immunol Lett 78:183-187
120. Zipp F, Weller M, Calabresi PA et al (1998) Increased serum levels of soluble CD95 (APO-1/Fas) in relapsing-remitting multiple sclerosis. Ann Neurol 43:116-120

121. Mu X, He J, Anderson DW, Trojanowski JQ, Springer JE (1996) Altered expression of bcl-2 and bax mRNA in amyotrophic lateral sclerosis spinal cord motor neurons. Ann Neurol 40:379-386
122. Su JH, Satou T, Anderson AJ, Cotman CW (1996) Up-regulation of Bcl-2 is associated with neuronal DNA damage in Alzheimer's disease. Neuroreport 7:437-440
123. MacGibbon GA, Lawlor PA, Sirimanne ES et al (1997) Bax expression in mammalian neurons undergoing apoptosis, and in Alzheimer's disease hippocampus. Brain Res 750:223-234
124. Nagy ZS, Esiri MM (1997) Apoptosis-related protein expression in the hippocampus in Alzheimer's disease. Neurobiol Aging 18:565-571
125. Tatton NA (2000) Increased caspase 3 and Bax immunoreactivity accompany nuclear GAPDH translocation and neuronal apoptosis in Parkinson's disease. Exp Neurol 166:29-43
126. Sjogren M, Andreasen N, Blennow K (2003) Advances in the detection of Alzheimer's disease-use of cerebrospinal fluid biomarkers. Clin Chim Acta 332:1-10
127. Sjogren M, Minthon L, Davidsson P et al (2000) CSF levels of tau, beta-amyloid(1-42) and GAP-43 in frontotemporal dementia, other types of dementia and normal aging. J Neural Transm 107:563-579
128. Kapaki E, Kilidireas K, Paraskevas GP et al (2001) Highly increased CSF tau protein and decreased beta-amyloid (1-42) in sporadic CJD: a discrimination from Alzheimer's disease? J Neurol Neurosurg Psychiat 71:401-403
129. Sjogren M, Davidsson P, Wallin A et al (2002) Decreased CSF-beta-amyloid 42 in Alzheimer's disease and amyotrophic lateral sclerosis may reflect mismetabolism of beta-amyloid induced by disparate mechanisms. Dement Geriatr Cogn Disord 13:112-118
130. Sjogren M, Gisslen M, Vanmechelen E, Blennow K (2001) Low cerebrospinal fluid beta-amyloid 42 in patients with acute bacterial meningitis and normalization after treatment. Neurosci Lett 314:33-36
131. Lewczuk P, Esselmann H, Otto M et al (2004) Neurochemical diagnosis of Alzheimer's dementia by CSF Abeta42, Abeta42/Abeta40 ratio and total tau. Neurobiol Aging 25:273-281
132. Kohira I, Tsuji T, Ishizu H et al (2000) Elevation of neuron-specific enolase in serum and cerebrospinal fluid of early stage Creutzfeldt-Jakob disease. Acta Neurol Scand 102:385-387
133. Kropp S, Zerr I, Schulz-Schaeffer WJ et al (1999) Increase of neuron-specific enolase in patients with Creutzfeldt-Jakob disease. Neurosci Lett 261:124-126
134. Blennow K, Wallin A, Ekman R (1994) Neuron specific enolase in cerebrospinal fluid: a biochemical marker for neuronal degeneration in dementia disorders? J Neural Transm Park Dis Dement Sect 8:183-191
135. Finsterer J, Exner M, Rumpold H (2004) Cerebrospinal fluid neuron-specific enolase in non-selected patients. Scand J Clin Lab Invest 64:553-558
136. Noppe M, Crols R, Andries D, Lowenthal A (1986) Determination in human cerebrospinal fluid of glial fibrillary acidic protein, S-100 and myelin basic protein as indices of non-specific or specific central nervous tissue pathology. Clin Chim Acta 155:143-150
137. Herrmann M, Vos P, Wunderlich MT et al (2000) Release of glial tissue-specific proteins after acute stroke: a comparative analysis of serum concentrations of protein S-100B and glial fibrillary acidic protein. Stroke 31:2670-2677
138. Ross SA, Cunningham RT, Johnston CF, Rowlands BJ (1996) Neuron-specific enolase as an aid to outcome prediction in head injury. Br J Neurosurg 10:471-476
139. Nylen K, Ost M, Csajbok LZ et al (2006) Increased serum-GFAP in patients with severe traumatic brain injury is related to outcome. J Neurol Sci 240:85-91

140. Fukuyama R, Izumoto T, Fushiki S (2001) The cerebrospinal fluid level of glial fibrillary acidic protein is increased in cerebrospinal fluid from Alzheimer's disease patients and correlates with severity of dementia. Eur Neurol 46:35-38
141. Wallin A, Blennow K, Rosengren LE (1996) Glial fibrillary acidic protein in the cerebrospinal fluid of patients with dementia. Dementia 7:267-272
142. Albrechtsen M, Sorensen PS, Gjerris F, Bock E (1985) High cerebrospinal fluid concentration of glial fibrillary acidic protein (GFAP) in patients with normal pressure hydrocephalus. J Neurol Sci 70:269-274
143. Malmestrom C, Haghighi S, Rosengren L et al (2003) Neurofilament light protein and glial fibrillary acidic protein as biological markers in MS. Neurology 61:1720-1725
144. Rosengren LE, Lycke J, Andersen O (1995) Glial fibrillary acidic protein in CSF of multiple sclerosis patients: relation to neurological deficit. J Neurol Sci 133:61-65
145. McShane LM, Altman DG, Sauerbrei W et al (2005) REporting recommendations for tumour MARKer prognostic studies (REMARK). Br J Cancer 93:387-391
146. Visser PJ, Scheltens P, Verhey FR (2005) Do MCI criteria in drug trials accurately identify subjects with predementia Alzheimer's disease? J Neurol Neurosurg Psychiat 76:1348-1354

[illegible]

DESIGNING MS TRIALS FOR NEURODEGENERATION

Chapter 16

Critical Review of Existing Trials

G. Comi

Introduction

Epidemiological studies indicate that about 90% of multiple sclerosis (MS) patients experience a progression of the disease at some time during their life [1, 2]. In about 15% of patients, the disease starts with an insidious onset followed by a progressive phase, with or without phases of stability; the so-called primary progressive (PP) MS. In all other patients, the early phase of MS is marked by acute attacks characterized by unifocal (2/3 of patients) or multifocal white matter disease [3]. Attacks are usually followed by complete or almost complete recovery; however, pathological studies demonstrate the presence of an axonal transection inside the lesion [4-6]. Axonal loss is predominant in lesions appearing in the early phases of the disease [7] and decreases over time. A large amount of damage occurs in areas with a considerable infiltration of T lymphocytes (especially CD8+ T cells) and macrophages [7], indicating a correlation between inflammation and axonal damage – a relationship which has also been shown by magnetic resonance imaging (MRI) studies. The acute axonal damage also occurs because of the products of inflammation, such as nitric oxide and tumor necrosis factor [8]. Recent studies indicate that a high level of electrical activity may increase the axonal degeneration of partially or completely demyelinated lesions at the nodes of Ranvier, as a consequence of the activation of glutamate receptors increasing the entry of calcium into the axon [9]. Recent studies suggest that axonal damage may also be independent of demyelination [10] and may be caused by antibodies against axonal antigens [11].

Axonal degeneration may be observed also in the so called normal appearing white matter, which could not be fully explained by a secondary degeneration of axons transacted inside the lesion [6]. Magnetization transfer studies demonstrated a decreased ratio in normal appearing white matter of patients in the early phases of the disease, increasing in magnitude with the disease evolution [12, 13].

As well as acute axonal damage, there is pathological [6], electrophysiological [14, 15], and MRI [16] evidence that a secondary axonal degeneration is also present in MS, both inside the lesions and in the normal-appearing white matter. The amount of this secondary degeneration, compared with the acute degeneration, is unknown. Many explanations have been proposed for secondary axonal degeneration. Naked demyelinated axons may be more susceptible to degeneration because they have lost the trophic support from the oligodendrocyte – a hypothesis supported by the observation that remyelinated axons are protected from further damage [7]. The secondary degeneration may also occur because repeated episodes of demyelination have exhausted the availability of oligodendrocyte precursor or decreased their remyelinating efficiency [17]. A primary pathology of oligodendrocytes could also explain the inefficient remyelination in some cases – an explanation which has been proposed for patients with PPMS [18]. Finally we should consider the extreme hypothesis that MS is a primary progressive degenerative disease with a secondary inflammatory response. Some recent pathological studies have indicated the key role of the axonal degeneration of sensory and motor pathways in determining irreversible disability in MS. The clinical observation that the topographic pattern of irreversible, progressive neurological deficits in MS depends on the localization of the previous attacks would seem to confirm this observation.

Clinical Trials in Progressive MS

Interferons

Five class I clinical trials investigated the effects of interferons (IFN) on disease progression in patients with secondary progressive (SP) MS. The first clinical trial was a multicenter, double-blind, placebo-controlled study involving 718 patients in 32 European centers [19]. The study demonstrated that IFNβ-1b at a dose of 8 MIU administered every other day by subcutaneous injection has a beneficial effect. The study was stopped by the External Advisory Board on the grounds of clinical efficacy before the scheduled 3-year period was completed. The findings of this study demonstrated a statistically significant reduction in relapse rate in patients receiving IFNβ-1b (about 30%). In addition, the proportion of patients who worsened was lower in those on therapy compared with those on placebo (38.9% and 49.7%, respectively); such findings were observed both in patients who complained of neurological attacks and in those who remained relapse-free during the treatment period. MRI data supported the clinical efficacy: both the MRI activity and the MRI lesion load were significantly reduced in the treated group compared with the placebo group [20].

A North-American study evaluated the efficacy and tolerability of 8 MIU and 5 MIU (mean dose 9.6 MIU) IFNβ1-b administered subcutaneously every other day, compared with placebo, in patients with SPMS. A sample group of 939 patients were randomly assigned to placebo or IFNβ1-b and were examined every 12 weeks for 3 years [21]. The results, considering the primary endpoint (time to progression of neurological impairment as defined by a confirmed increase of at least 1 EDSS point over baseline or 0.5 EDSS point if baseline EDSS is 6.0 or greater), demonstrated the same trend for both IFN-treated and placebo-treated patients. A positive result was found only in the subgroup of SP patients with superimposed relapses. IFNβ-1b treatment resulted in improvement on secondary outcome measures involving clinical relapses, newly active MRI lesions, and accumulated burden of disease on T2-weighted MRI.

The Secondary Progressive Efficacy Clinical Trial of Recombinant Interferon-beta-1a in MS (SPECTRIMS) study, which explored the effects of IFNβ-1a (Rebif 44 mcg taken three times a week) also failed to show a statistically significant slowing of progression, although a significant reduction in relapse rate (0.71 in the placebo arm vs. 0.50 in the active arm) was achieved [22].

MRI analysis demonstrated a significant reduction in the median number of active lesions and in the accumulation of T2 lesion burden [23]. Significant treatment effects were also seen on other exacerbation-related outcomes and on a composite measure incorporating five separate clinical and MRI outcomes. Post-hoc analysis showed a greater benefit in women and in patients with one or more relapses in the 2 years prior to the study.

In the fourth study, the efficacy of 30 μg weekly IFNβ-1a (Avonex) was tested in a multicenter North-American trial performed with 436 SPMS patients [24]. After 2 years there was a significant decrease in the mean Multiple Sclerosis Functional Composite Score (MSFC) in the treated group vs. the placebo group, mostly explained by positive effects in the "nine-hole peg test." The mean EDSS and the proportion of patients with significant EDSS progression were not significantly different in the two groups; however, some items from the quality of life scale also favored the active treatment. The number of active MRI lesions was significantly reduced at year 1 and year 2 in the IFNβ-1a arm.

In the recently published northern European multicenter study, a low dose of IFNβ-1a (Rebif 22 μg subcutaneous, administered weekly) was tested in a double-blind, placebo-controlled study in 371 SPMS patients [25]. The duration of the clinical trial was 3 years, with clinical assessments every 6 months. There was no treatment effect on the time to confirmed progression of EDSS; moreover, there was also no effect on relapse rate, which was very low in both arms: 0.27 in the placebo arm vs. 0.25 in the active arm. The brain MRI was not performed. These results are not surprising because this low dose of IFNβ-1a is also not effective in RRMS patients [26].

The results of studies performed on SPMS patients indicate that treatment with IFNβ maintains the same magnitude of effects on disease activity observed in relapsing-remitting (RR) MS patients, with a very homogeneous behavior across studies (Fig. 1); on the other hand, the effects on disability progression are modest and are seen only in the European IFNβ-1b study. Less-severely disabled patients with clinical or MRI evidence of disease activity seem to benefit more than more-severely disabled patients with slow progression of disability. However, the patient baseline characteristics were different in the different studies. The comparison between the North-American and European IFNβ-1b studies is of particular interest because of the same active treatment and the same modality of assessment. The patients in the European IFNβ-1b study were

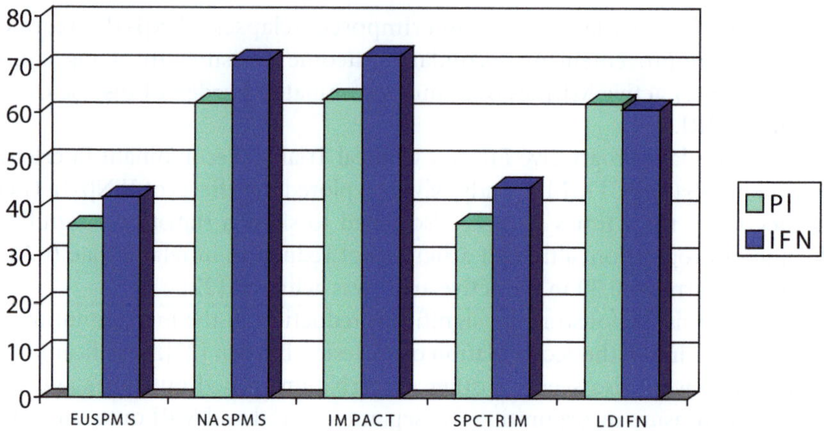

Fig. 1. Percentage of patients free from attacks in the active arm and in the placebo arm. For the SPECTRIM trial the data of the high dose arm only are presented. All the studies except the low dose clinical trial resulted in an higher proportion of patients free from relapses in the active arm compared to placebo arm

younger, had a slightly shorter duration of disease, a higher relapse rate in the 2 years prior to entry and, in the frequent MRI subgroup, a greater proportion had contrast-enhancing lesions [27]. This higher disease activity in the European study is the probable explanation for the positive effects observed on disease progression. The patient baseline characteristics in the SPECTRIMS study of subcutaneous IFNβ-1a, which also failed to show a significant treatment effect on EDSS progression, were somewhere between the North-American and the European IFNβ-1b studies, while the International MS Secondary Progressive Avonex Controlled Trial (IMPACT) study of intramuscular IFNβ-1a [24] recruited patients similar to those taking part in the North-American IFNβ-1b study.

Adverse effects related to IFN are more frequent in SPMS patients – for instance, the increase in spasticity – and may be quite disabling; therefore the decision to start treatment should be taken with great caution.

Two phase-II clinical trials have been performed in PPMS patients. A first exploratory double-blind, placebo-controlled, randomized study failed to show effects of IFNβ-1a on both clinical and MRI parameters [28]. A second more recent clinical trial in PPMS and transitional MS patients recruited 73 patients randomized either to IFNβ-1b every other day or placebo for a 2-year study. Time to confirmed progression was not significantly different in the two arms; however, a trend in favor of the active treatment emerged. A significant difference favoring IFNβ-1b treatment was indeed observed on the MSFC score. Furthermore, the active treatment significantly reduced the accumulation of T2 and T1 lesion load and the number of new T2 lesions [29]. At present there is no evidence from class I trials of the efficacy of IFNβ in patients with PPMS, so the treatment is not recommended.

Glatiramer Acetate (GA)

During the 1980s, a large body of data derived from small open-label and phase-II trials suggested the efficacy of GA in reducing the frequency of relapses in RRMS patients [30, 31]. A North-American double-blind, randomized, controlled-versus-placebo trial confirmed the efficacy of GA on relapse rate [32]. Subsequently the European-Canadian clinical trial demonstrated the efficacy of GA in reducing the number and volume of gadolinium-enhancing lesions and the accumulation of T2 lesion load [33]. The effect of GA on SPMS patients was explored in a small controlled study. Results suggested the efficacy of the drug in reducing disability progression in SPMS patients [34]. Unfortunately no other clinical trials have been performed in SPMS patients. More recently, a large multicenter, placebo-controlled clinical trial testing the efficacy of GA in PPMS has been interrupted after the second interim analysis because of the results of the futility analysis.

Immunoglobulins

There is controversial and conflicting evidence on the effects of high-dose intravenous immunoglobulins (IVIg) in the treatment of patients with RRMS [35]. The efficacy of IVIg in SPMS patients has been tested in a phase-III, double-blind, placebo-controlled, randomized clinical trial involving 318 patients randomly assigned to 1 g/kg IVIg per month, or an equivalent volume of placebo, for 27 months. No beneficial effects were observed on the primary endpoint, the time to confirmed progression of disability, or on any of the other clinical or MRI parameters [36, 37]. In an ancillary study, the histogram analysis of the magnetization transfer ratio of the normal-appearing brain tissue showed a more severe progression of the damage in the placebo group compared with the IVIg-treated group [38].

Immunosuppressive Treatments

The effects of immunosuppression have been tested in the past in many small and frequently uncontrolled trials and, more recently, in some better designed studies. Cyclophosphamide is an alkylating agent used to treat malignancies and immune-mediated diseases. About 30 years ago the efficacy of cyclophosphamide was tested in open uncontrolled studies at some European MS centers [39, 40]. The first randomized controlled trial to test cyclophosphamide in 20 patients was reported by Hauser et al. [41]. Two placebo-controlled studies reported that no positive clinical effects were observed with a 2-week induction regimen [42, 43]. The Northeast Cooperative Multiple Sclerosis Treatment Group demonstrated that booster therapy every 2 months significantly reduced the progression of disability compared with the simple induction regimen [44]. Moreover, the effects were found in younger patients only. Some recent small open studies recently confirmed the role of cyclophosphamide in dramatically reducing disease activity, either alone [45, 46] or in combination with immunomodulating agents [47]. In this latter study, the frequency of treatment failure was significantly higher in patients treated with monthly methylprednisolone compared with patients treated with a monthly combination of methylprednisolone and cyclophosphamide.

Mitoxantrone is an anthracycline derivative which achieves its cytotoxicity by arresting the cell cycle at the G2-M and S interphase. The immunosuppressive effect involves B and T lymphocytes and macrophages [48, 49]. A series of controlled and uncontrolled studies has demonstrated that the drug dramatically reduces the clinical and MRI activity in RRMS and SPMS patients [50-53]. In a double-blind trial, 49 patients with relapsing SPMS were randomized to receive either 13 infusions of mitoxantrone (12 mg/m^2) or methylprednisolone (1 g). At years 1 and 2 there was a significant reduction in the proportion of patients with disease progression in the mitoxantrone-treated group compared with the placebo group [50]. The difference was not statistically significant at year 3, probably because of the small sample size. Some clinical and MRI secondary endpoints also favored the active treatment. In the multicenter French and British open trial, patients with active MS were randomized to receive monthly intravenous injections of 20 mg of mitoxantrone plus 1 g of methylprednisolone or methylprednisolone alone for 6 months, with frequent brain MRI evaluation [51]. The mean number of new enhancing lesions during the 6-month period was significantly reduced and the proportion of patients with inactive scans was significantly increased in the mitoxantrone group compared with the placebo group. In the Mitoxantrone in Multiple Sclerosis (MIMS) trial, patients were randomized to receive 12 mg/m^2 or 5 mg/m^2 of mitoxantrone or placebo every 3 months for 2 years [53]. The high dose significantly reduced the relapse rate and the progression of disability. Brain MRI was performed in a subgroup of patients only. Results were less convincing: the primary endpoint, the number of gadolinium-enhancing lesions at months 12 and 24, was not met [54]. The number of new T2 lesions at 24 months was 1.94 in the placebo arm but dropped significantly to 0.29 in the mitoxantrone high-dose arm.

As stated in two recent reviews [55, 56], there is some evidence of a moderate effect of mitoxantrone in reducing disability progression and the frequency of exacerbations in patients with RRMS, SPMS, or progressive relapsing MS. Given the risks of cardiotoxicity [57] and leukemia, treatment with mitoxantrone should be limited to the non-responders to immunomodulatory treatments who are in an active phase of the disease.

Conclusions

There is converging evidence from experimental studies, clinical and pathological findings, and clinical trials that inflammation is the predominant cause of axonal loss in the early phases of MS. The reduction in relapse rate and accumulation of new lesions in patients with clinically isolated syndrome (CIS) and RRMS is associated with a modest progression of disability and to a decrease in brain atrophy. Conversely, in SPMS patients most of the nervous damage is due to mechanisms of degeneration, which are independent from the ongoing inflammation. Clinical trials performed in PPMS and SPMS demonstrate that the reduction of inflammation is similar to that obtained in CIS and RR patients; however, the positive effects on the accumulation of disability are definitely less relevant. From a practical point of view, in SPMS patients with clinical or MRI evidence of disease activity, both IFN and immunosuppressive drugs may produce benefits. The different safety profiles of the two classes of treatment suggest that it is wise to start with IFNs; however, in patients with very active disease or with intolerance to IFNs, the possibility of using immunosuppressive drugs should be considered. Limited effects of disease-modifying agents in progressive disease courses strongly support the necessity of exploring the value of neuroprotective and neurotrophic treatments as a useful strategy to protect axons from damage and death.

References

1. Weinshenker BG, Bass B, Rice GHPA et al (1989) The natural history of multiple sclerosis: a geographical based study 1. Clinical course and disability. Brain 112:133-146
2. Runmarker B, Andersen O (1993) Prognostic factors in a multiple sclerosis incidence cohort with twenty-five years of follow-up. Brain 116:117-134
3. Wingerchuk DM, Weinshenker BG (2000) Multiple sclerosis: epidemiology, genetic, classification, natural history and clinical outcome measures. Neuroimaging Clin N Am 10:611-623
4. Ferguson B, Matyszak MK, Esiri MM, Perry VH (1997) Axonal damage in acute multiple sclerosis lesions. Brain 120:393-399
5. Trapp BD, Peterson J, Ransohoff RM et al (1998) Axonal transection in the lesions of multiple sclerosis. New Engl J Med 338:278-285
6. Kornek B, Storech MK, Weissert R et al (2000) Multiple sclerosis and chronic autoimmune encephalomyelitis. Am J Pathol 157:267-276
7. Kuhlmann T, Lingfeld G, Bitsch A et al (2002) Acute axonal damage in multiple sclerosis is most extensive in early disease stages and decreases over time. Brain 125:2202-2212
8. Smith KJ, Kapoor R, Felts PA (2001) Electrically active axons degenerate when exposed to nitric oxide. Ann Neurol 49:470-476
9. Kapoor R, Davied M, Blaker PA et al (2003) Blockers of sodium and calcium entry protect axons from nitric oxide-mediated degeneration. Ann Neurol 53:174-180
10. Bitsch A, Schuchardt J, Bunkowski S et al (2000) Acute axonal injury in multiple sclerosis: correlation with demyelination and inflammation. Brain 123:1174-1183
11. Raine C, Cannella B, Hauser S, Genain C (1999) Demyelination in primate autoimmune encephalomyelitis and acute multiple sclerosis lesions: a case for antigen-specific antibody mediation. Ann Neurol 46:144-160
12. Filippi M, Inglese M, Rovaris M et al (2000) Magnetization transfer imaging to monitor the evolution of MS: a 1-year follow-up study. Neurology 55:940-946
13. Iannucci G, Tortorella C, Rovaris M et al (2001) Prognostic value of MR and magnetization transfer imaging findings in patients with clinically isolated syndromes suggestive of multiple sclerosis at presentation. Am J Neuroradiol 21:1034-1038

14. Sater RA, Rostami AM, Galetta S et al (1999) Serial evoked potential studies and MRI imaging in chronic progressive multiple sclerosis. J Neurol Sci 171(2):79-83
15. Leocani L, Comi G (2000) Neurophysiological investigations in multiple sclerosis. Curr Opin Neurol 13:255-261
16. Rocca MA, Mastronardo G, Rodegher M et al (1999) Long-term changes of magnetization transfer-derived measures from patients with relapsing-remitting and secondary progressive multiple sclerosis. Am J Neuroradiol 20:821-827
17. Linington C, Engelhardt B, Kapocs G, Lassmann H (1992) Induction of persistently demyelinated lesions in the rat following the repeated adoptive transfer of encephalitogenic T cells and demyelinating antibody. J Neuroimmunol 40:219-224
18. Lucchinetti C, Bruck W, Parisi J et al (1999) A quantitative analysis of oligodendrocytes in multiple sclerosis lesions: a study of 113 cases. Brain 122:2279-2295
19. European Study Group on Interferon β-1b in Secondary Progressive MS (1998) Placebo-controlled multicentre randomised trial of interferon β-1b in treatment of secondary progressive multiple sclerosis. Lancet 352:1491-1497
20. Miller DH, Molyneux PD, Barker GJ et al (1999) Effect of interferon β-1b on magnetic resonance imaging outcomes in secondary progressive multiple sclerosis: results of a European multicenter, randomized, double-blind, placebo-controlled trial. Ann Neurol 46:850-859
21. North American Study Group on Interferon beta-1b in Secondary Progressive MS (2004) Interferon beta-1b in secondary progressive MS: results from a 3-year controlled study. Neurology 63:1788-1795
22. Secondary Progressive Efficacy Clinical Trial of Recombinant Interferon-beta-1a in MS (SPECTRIMS) Study Group (2001) Randomized controlled trial of interferon-beta-1a in secondary progressive MS: clinical results. Neurology 56:1496-1504
23. Li DKB, Zhao GJ, Paty DW (2001) Randomized controlled trial of interferon-beta-1a in secondary progressive MS: MRI results. Neurology 56:1505-1513
24. Cohen JA, Cutter GR, Fischer JS et al (2002) Benefit of interferon β-1a on MSFC progression in secondary progressive MS. Neurology 59:679-687
25. Andersen O, Elovaara I, Färkkilä M et al (2004) Multicentre, randomised, double blind, placebo controlled, phase III study of weekly, low dose, subcutaneous interferon beta-1a in secondary progressive multiple sclerosis. J Neurol Neurosurg Psychiat 75:706-710
26. The Once Weekly Interferon for MS Study Group (OWIMS) (1999) Evidence of interferon β-1a dose response in relapsing-remitting MS: The OWIMS Study. Neurology 53:679-686
27. Kappos L, Weinshenker B, Pozzilli C et al (2004) Interferon beta-1b in secondary progressive MS: a combined analysis of the two trials. Neurology 63:1779-1787
28. Leary SM, Thomson AJ (2003) Interferon beta-1a in primary progressive multiple sclerosis. J Neurol Sci 206:215-216
29. Montalban X, Thomson AJ (2002) Workshop on primary progressive multiple sclerosis: meeting summary. Mult Scler 8:177-178
30. Bornstein MB, Miller A, Teitelbaum D et al (1982) Multiple sclerosis: trial of a synthetic polypeptide. Ann Neurol 11:317-319
31. Bornstein MB, Miller A, Slagle S et al (1987) A pilot trial of COP 1 in exacerbating-remitting multiple sclerosis. New Engl J Med 317:408-414
32. Johnson KP, Brooks BR, Cohen JA et al and Copolymer 1 Multiple Sclerosis Study Group (1995) Copolymer 1 reduces relapse rate and improves disability in relapsing-remitting multiple sclerosis: results of the phase III multicenter, double-blind placebo-controlled trial. Neurology 45:1296-1276
33. Comi G, Filippi M, Wolinsky JS and the European-Canadian Glatiramer Acetate Study Group (2001) The European/Canadian multicenter, double-blind, random-

ized, placebo-controlled study of the effects of glatiramer acetate on magnetic resonance imaging-measured disease activity and burden in patients with relapsing-remitting multiple sclerosis. Ann Neurol 49:290-297
34. Bornstein MB, Miller A, Slagle S et al (1991) A placebo-controlled, double-blind, randomized, two-center, pilot trial of Cop 1 in chronic progressive multiple sclerosis. Neurology 41:533-539
35. Sorensen PS, Fazekas F, Lee M (2002) Intravenous immunoglobulin G for the treatment of relapsing-remitting multiple sclerosis: a meta-analysis. Eur J Neurol 9:557-563
36. Hommes OR, Sorensen PS, Fazekas F et al (2004) Intravenous immunoglobulin in secondary progressive multiple sclerosis: randomised placebo-controlled trial. Lancet 364:1149-1156
37. Fazekas F, Sorensen PS, Filippi M et al (2005) MRI results from the European Study on Intravenous Immunoglobulin in Secondary Progressive Multiple Sclerosis (ESIMS). Mult Scler 11:433-440
38. Filippi M, Rocca MA, Pagani E et al (2004) European study on intravenous immunoglobulin in multiple sclerosis: results of magnetization transfer magnetic resonance imaging analysis. Arch Neurol 61:1409-1412
39. Hommes OR, Prick JJG, Lamers KJB (1975) Treatment of the chronic progressive form of multiple sclerosis with a combination of cyclophosphamide and prednisone. Clin Neurol Neurosurg 78:59-73
40. Gonsette R, Demonty L, Delmotte P (1977) Intensive immunosuppression with cyclophosphamide in remittent forms of multiple sclerosis: follow up of 110 patients for 2-6 years. J Neuroimmunol 214:173-181
41. Hauser SL, Dawson DM, Lehrich JR et al (1983) Intensive immunosuppression in progressive multiple sclerosis: a randomized three-arm study of high-dose intravenous cyclophosphamide, plasma exchange, and ACTH. N Engl J Med 308:173-180
42. Canadian Cooperative Study Group (1991) The Canadian Cooperative trial of cyclophosphamide and plasma exchange in progressive multiple sclerosis. Lancet 337:441-446
43. Likowsky VH, Fireman B, Elmore R et al (1991) Intensive immunosuppression in chronic progressive multiple sclerosis: the Kaiser study. J Neurol Neurosurg Psychiat 54:1055-1060
44. Weiner HL, Mackin GA, Orav EJ et al (1993) Intermittent cyclophosphamide pulse therapy in progressive multiple sclerosis: final report of the Northeast Cooperative Multiple Sclerosis Treatment Group. Neurology 43:910-918
45. Gobbini MI, Smith ME, Richert ND et al (1999) Effect of open-label pulse cyclophosphamide therapy on MRI measures of disease activity in five patients with refractory relapsing-remitting multiple sclerosis. J Neuroimmunol 99:142-149
46. Perini P, Gallo P (2003) Cyclophosphamide is effective in stabilizing rapidly deteriorating SPMS. J Neurol 250:834-838
47. Smith DR, Weinstock-Guttman B, Cohen JA et al (2005) A randomized blinded trial of combination therapy with cyclophosphamide in patients with active multiple sclerosis on interferon beta. Mult Scler 11:573-582
48. Fidler JM, Dejoy SQ, Gibbons JJ Jr (1986) Selective immunomodulation by the antineoplastic agent mitoxantrone I. Suppression of B lymphocyte function. J Immunol 137:727-732
49. Fidler JM, Dejoy SQ, Smith FR 3rd, Gibbons JJ Jr (1986) Selective immunomodulation by the antineoplastic agent mitoxantrone II. Nonspecific adherent suppressor cells derived from mitoxantrone. J Immunol 136:2747-2754
50. Millefiorini E, Gasperini C, Pozzilli C et al (1997) Randomized placebo-controlled trial of mitoxantrone in relapsing-remitting multiple sclerosis: 24 month clinical and MRI outcome. J Neurol 244:153-59

51. Edan G, Miller D et al (1995) Evaluation of the efficacy of mitoxantrone by use of MRI: a multicenter randomized study in multiple sclerosis. Neurology 242(Suppl 2): S38 [abstract]
52. Hartung HP, Gonsette R, Kwiecinski H et al (2002) Mitoxantrone in progressive multiple sclerosis: a placebo-controlled, double-blind, randomised, multicentre trial. Lancet 360(9350):2018-2025
53. van de Wyngaert FA, Beguin C, D'Hooghe MB et al (2001) A double blind clinical trial of mitoxantrone versus methylprednisolone in relapsing secondary progressive multiple sclerosis. Acta Neurol Belg 244:153-159
54. Krapf H, Morrissey SP, Zenker O et al (2005) Effect of mitoxantrone on MRI in progressive MS: results of the MIMS trial. Neurology 65:690-695
55. Martinelli Boneschi F, Rovaris M, Capra R, Comi G (2005) Mitoxantrone for multiple sclerosis. Cochrane Database of Systematic Reviews 4:CD002127
56. Fox EJ (2006) Management of worsening multiple sclerosis with mitoxantrone: a review. Clin Therapeut 28:461-474
57. Jammohammed R, Milligan DW (1989) Mitoxantrone induced congestive heart failure in patients previously treated with anthracyclines. Br J Haematol 71:292-293

Chapter 17

Design for the Next Trials of Neurodegeneration

P. SOELBERG SØRENSEN

Introduction

During the past decade, several new medications have become available for the treatment of multiple sclerosis (MS). The main effect of these therapies, whether they are immunomodulatory or immunosuppressive types of treatment, has been their strong impact on the inflammatory component of the MS disease process. These anti-inflammatory therapies have, with various degrees of success, suppressed signs of inflammation as detected by magnetic resonance imaging (MRI) and have reduced the incidence of the clinical correlate to inflammation, i.e., the acute relapses [1-7], although they have had little effect on disease progression caused by permanent demyelination and axonal loss. Hence, it has been hypothesized that there are two different disease mechanisms at work in patients with MS: inflammation and neurodegeneration. The neurodegenerative disease process has been associated with the progressive phases of MS, either primary progressive (PP) MS or secondary progressive (SP) MS, during which the disease activity is thought to be driven by degenerative rather than inflammatory processes. However, recent studies have emphasized the presence of axonal damage early in the course of the disease [8], a fact that was already known by Charcot, as shown in his seminal studies of MS [9]. According to one theory, both inflammatory and degenerative disease processes are present from the onset of the disease and proceed independently; with inflammation being most prominent in the early phases of the disease, whereas neurodegeneration dominates during the later stages [10]. On the other hand, there are arguments for inflammation being the culprit for both acute relapses and disease progression [11-13]. Recent studies have indeed shown that inflammation is also prominent in patients with PPMS and SPMS, although this inflammation is thought to be compartmentalized within the CNS and is independent of the activation of T-cells in the peripheral bloodstream [14]. This low-burning, widespread inflammation is thought to be mediated by activated microglia that interact with T- and B-cells from infiltrates in the meninges and in the perivascular spaces, causing demyelination and axonal injury in plaques as well as diffusely in the white and gray matter of the brain [15, 16].

Whereas relapses and relapse-related disability are the clinical correlates to the formation of inflammatory demyelinating plaques in MS, which are indicated on MRI scans by gadolinium-positive lesions and new T2 lesions [17], the long-term disability and cognitive changes are considered to be clinical correlates of permanent demyelination and axonal and neuronal loss, which are shown on MRI scans as T1 black holes and atrophy [18-21].

None of the existing MS drugs have convincingly been shown to prevent disease progression, which would have suggested that they provide either neuroprotection or repair. However, a number of compounds are claimed to possess neuroprotective properties, and there are plans in place to test these drugs in clinical trials during the coming years.

Previous Trials

None of the large-scale trials of disease-modifying drugs have had long-term disease progression as the primary endpoint [1-7]. A few trials have used 2- or 3-year progression on the Expanded Disability Status Scale (EDSS) as the primary endpoint, but increases in EDSS in short-term trials in relapsing/remitting (RR) MS are merely brought about by incomplete recovery after relapses rather than by disease progression unrelated to relapses. Trials in PPMS and SPMS have been unsuccessful regarding the prevention of long-term disability measured using EDSS, and changes in cognitive functions have not been used as primary or secondary endpoints in large-scale trials [22-26]. Furthermore, analysis of the negative results of trials in PPMS and SPMS has shown that the use of disease progression as primary endpoint requires very large patient populations monitored over long periods of time, and hence there is an urgent need for MRI endpoints as surrogates for clinical disease progression and neurodegeneration.

MRI to Provide Surrogate Endpoints for Neurodegeneration

Conventional MRI Techniques

The number of new or enlarging lesions on T2-weighted MRI scans has traditionally been used to measure disease activity in the course of a clinical trial, and the evolution of T2 lesion area or volume (burden of disease) has been used as a surrogate for short-term disease progression. However, T2 lesions lack neuropathological specificity and are unable to provide any distinction between edema, demyelination, remyelination, gliosis, or axonal loss; and T2-weighted images are unable to demonstrate pathology in normal-appearing brain tissue [27]. Hence, T2-weighted images are not well suited for measuring neurodegeneration.

T1-weighted images obtained after the administration of gadolinium are usually used in short-term trials with frequent scanning as a surrogate endpoint of acute plaque activity related to relapses, and, like T2 lesions, neither the measurement of new gadolinium-positive lesions or the evolution of gadolinium lesion area or volume are suitable for imaging neurodegeneration. Hypointense lesions on T1-weighted images (black holes) are thought to reflect permanent tissue loss, and the number of new T1 lesions or the evolution of T1 lesion area or volume are more appropriate MRI outcome measures in trials of neurodegeneration than changes in abnormalities on T2-weighted images [28]. Post-mortem MRI studies have shown that T1 black holes are associated with severe tissue destruction, including axonal loss [20].

The evolution of gadolinium-positive lesions observed with frequent MRI scans might be appropriate for studying neurodegeneration and repair in short-term trials. The frequency of new gadolinium-enhancing lesions that progress into T1 black holes can be studied with frequent MRI scans that allow the new gadolinium-positive lesions to be followed up for 6 months [29]. The percentage of gadolinium-enhancing lesions that evolve into T1 black holes could be used in short-term trials as an MRI surrogate for neurodegeneration or, probably more appropriately, as a measure of failure of repair [30].

With conventional MRI techniques, measurement of tissue loss is the most frequently applied MRI surrogate for neurodegeneration. When tissue destruction occurs, the central nervous system responds with shrinkage, leading to a reduced volume of the brain parenchyma and enlargement of the cerebrospinal fluid spaces. There is no gold standard for measuring atrophy. Several MRI measures have been used to assess brain atrophy [18]. These include increased volume of the ventricles or of all cerebrospinal fluid (CSF) spaces, decreased volume of the total brain parenchymal structures, atrophy of the corpus callosum, and cortical atrophy. The measures are often normalized to a standard intracranial volume. Different techniques, such as magnetization-prepared rapid gradient-echo (MP-RAGE) sequences, can yield 2D or 3D images of atrophy [18].

The most commonly applied measure is the brain parenchymal fraction [31], which is a measure of whole-brain atrophy defined as the ratio of the brain parenchymal volume to the total volume within the brain surface contour. This measure can be obtained using semi-automated methods [18].

There are several confounding factors known to influence measurements of the cerebral volume, including inflammation-derived edema, dehydration, intracranial hypertension, alcohol intake, and steroid therapy, the latter being very important in patients with RRMS. Measurements of brain atrophy should not be performed within the 2-3 months after steroid therapy [32]. Furthermore, the initiation of a strong anti-inflammatory therapy will tend to diminish inflammation and edema and thereby reduce brain volume. This may lead to an erroneous finding of accelerated brain tissue loss after the initiation of anti-inflammatory drugs [31]. Hence, when such drugs are used in clinical trials in which brain atrophy is an outcome measure, the baseline scanning should be postponed until 3 (or even 6) months after the initiation of therapy.

Measurements of spinal cord atrophy have traditionally been performed at C2 level. This measure might be the most appropriate MRI measure of neurodegeneration in progressive forms of MS, particularly in PPMS, where MRI changes of the brain are less prominent [33].

Large cross-sectional and longitudinal studies have shown that normal aging rates of global brain atrophy typically increase from an annual rate of 0.2% per year at 30-50 years of age to 0.3-0.5% per year at age 70-80 years, whereas MS patients typically have annual atrophy rates in the range 0.6-2.0%, with fairly similar rates being seen among RRMS, SPMS, and PPMS patients [33].

Non-conventional (New) MR Techniques

Non-conventional MR techniques include quantification of magnetization transfer ratio (MTR), diffusion tensor (DT) MRI, proton magnetic resonance spectroscopy (1H-MRS), and functional MRI (fMRI). These more advanced methods have increased our understanding of the pathogenesis of MS, but they are not all generally available and have not been standardized.

MT MRI is a new technique based on interaction and exchange between mobile protons in a free water pool with those bound to macromolecules. A low MTR indicates a reduced ability of macromolecules in tissue to exchange magnetization with surrounding water molecules, and is interpreted as an indication of damage to myelin and other cellular structures. MTR data are typically expressed as histograms and can be calculated throughout the whole brain, within lesions, or in normal-appearing white matter [34]. MTR is generally nonspecific and is influenced by edema, gliosis, and inflammation. However, MTR decreases are more profound in patients with progressive disease than in patients with RRMS, and brain MTR has predictive value for long-term development of disease progression [35-37]. Hence, MTR might be of value in long-term measurements of neurodegeneration. MTR is the first of the non-conventional MR techniques that have been applied in multicenter studies. In a European study of intravenous immunoglobulin in SPMS (ESIMS) there was a decrease in whole-brain MTR from baseline to month 24 in parallel with an increase in EDSS score [24, 38]. Studies have shown that treatment with interferon-beta (IFNβ) improves recovery of MTR in newly formed lesions compared with placebo, indicating that IFNβ affects lesion evolution as defined by MTR [39]. As MTR can be used to obtain information about structural changes, both within visible T2 lesions and in normal-appearing white matter, MTR might be of value in trials of neurodegeneration [40].

DT MRI can provide quantitative measures of water molecular motion. Using DT MRI, it is possible to measure the mean diffusivity and anisotropy, which display the structural integrity of the tissue and the degree of structural alignment within fiber tracts, respectively [41]. The mean apparent diffusivity in the normal-appearing brain tissue has been shown to increase over 1 year in patients with PPMS [42]. Hence, it seems that a demonstration of longitudinal changes in diffusivity could be used for the measurement of neurodegeneration. There are, however, no studies evaluating the use of DT MRI in clinical trials.

^{1}H-MRS allows the measurement of the concentration of the metabolite N-acetylaspartate (NAA) and has been used to measure neuronal integrity. NAA is a metabolite that is found almost exclusively in neurons and neuronal processes in the normal adult brain and is thought to express neuronal integrity or neuronal function. NAA decreases have been demonstrated in MS lesions, cortical gray matter, and whole brain of MS patients [43, 44]. In an acute plaque, NAA signal intensity is decreased and may stay decreased or show partial recovery lasting for several months [45]. NAA measurements in normal-appearing white matter are reduced in patients with early RRMS and show a further decrease in patients with progressive MS [46, 47]. 1H-MRS measurement of NAA in the whole brain has been developed and has shown that substantial axonal pathology exists even in patients in the earliest clinical phase of MS [48]. Recently it has been shown that NAA/creatinine ratios correlate with cognitive dysfunction [49]. In a small group of patients treated with IFNβ-1b, NAA/creatinine ratios were increased after 1 year of treatment [50], whereas other studies did not confirm this observation [51, 52]. Potentially, NAA could be of value in trials of neurodegeneration, but the small sizes of the patient cohorts studied and lack of control groups make studies of larger cohorts of patients necessary before the role of 1H-MRS in trials of neurodegeneration can be established. Although the technical demands of spectroscopy have until now limited therapeutic trials to single centers, it seems that multicenter trials may be feasible.

Functional MRI (fMRI) provides information about the neuronal mechanisms that underlie CNS function. fMRI can be used to study the extent and nature of brain plasticity following MS-related structural injury, with the potential to limit the clinical consequences of MS structural damage. When patients with clinically isolated syndrome (CIS) are observed performing simple motor tasks, an abnormal pattern of cortical activations characterized by increased activation in areas outside the motor cortex, both in the homo- and contra-lateral hemisphere, can be seen. It has been suggested that these changes in brain activation represent adapted cortical reorganization which is aimed at compensating for existing deficits [53]. A progressive exhaustion or inefficiency of the adaptive properties of the cerebral cortex could be among the factors responsible for the worsening of clinical disability in patients with MS [54, 55]. Hence, fMRI might be a potential tool for the study of neurodegeneration or repair processes. However, although spontaneous brain reorganization has been observed in MS patients, studies on therapeutic interventions and induced reorganizations are still rare and the results are controversial.

18-fluorodeoxyglucose positron emission tomography (18-FDG-PET) can be utilized to measure the cortical cerebral metabolic rate of glucose. A coupling between cerebral metabolic rates and neuroactivity has been shown in humans, which allows PET measurements to be used as an indirect estimate of brain function. Global cortical cerebral metabolism is decreased in MS patients compared with healthy controls, as well as the cortical metabolism in several regions of the brain [56]. A longitudinal study of cerebral glucose metabolism showed that the global cerebral metabolism decreases over time in progressive MS patients [57]. The global metabolic rate of glucose is correlated with cognitive dysfunction and with the T2 lesion load [56]. PET measurements of the cortical metabolic rate of glucose might be a valuable method for the study of neurodegeneration. However, PET studies are time-consuming, unpleasant for the patient, and incorporate the administration of radioactive isotopes, all of which make the method less applicable for repeated studies in MS.

Future Trials of Neurodegeneration

Trial Design and Patient Population

Trials of a new neuroprotective agent could include double-blind, placebo-controlled, parallel-group trials of the drug as mono-therapy or as an add-on to an existing approved treatment. Alternatively the trial could be a comparative trial to an existing treatment. The use of a virtual placebo group might be a possibility that has not yet been utilized [58, 59].

As neurodegeneration is already present in the early phases of MS, the selection of patients for a trial could include both patients with RRMS and patients with progressive MS.

Outcome Measures

Clinical endpoints are based on progression of permanent symptoms over time. Traditionally progression of 1 EDDS point (1½ points if EDSS is 0 and ½ point if EDSS is 6 or more) sustained for 3 (or preferably 6) months, has been used. Due to the day-to-day variation in patients' motor function and inter-examiner variability of EDSS ratings, a sustained progression of 2 EDSS points is considered a more robust endpoint. However, as fewer patients will progress 2 points during the study time, this endpoint requires larger patient populations. Measures of sustained progression can also include time to progression or proportion of patients progressing. Other disability-related endpoints are time to progression to a predefined EDSS score or proportion of patients progressing to a predefined EDSS score [58]. In an RRMS patient population, progressions to either EDSS 4 or EDSS 6 are appropriate endpoints, whereas in populations of patients with progressive MS, EDSS 6 or EDSS 7 can be used. Mean progression over time has been used as an outcome measure, but this measure is less feasible due to the inherent nonlinear properties of the EDSS. Recently, a new measure of disease progression has been proposed; the Multiple Sclerosis Severity Score [60].

The duration of trials of neurodegeneration has to be longer than trials of inflammatory activity [58]. Trials of RRMS should have a duration of at least

2 years, and preferably 3 years. However, it should be noted that EDSS progression in RRMS is a measure of incomplete recovery of relapses rather than one of gradual deterioration caused by neurodegeneration. Trials in progressive MS, and in particular in PPMS, need to have a duration of at least 3 years, and preferably 4 years, in order to attain sufficient power to show differences between placebo and intervention, due to the slow speed of progression, particularly in PPMS. Hence, such trials are very difficult to perform as placebo-controlled trials. Trials of a new drug with possible effects on neurodegeneration can be carried out either as trials using a virtual placebo group based on the natural history of patients with progressive MS or can be tested in add-on trials in which either the new drug or placebo is added to an established therapy. A common problem in add-on trials is the sample size calculation, because the disease progression in patients treated with the approved therapy is often unknown. In general, large patient populations are required in order to show a difference in the effect of the combination therapy versus the approved therapy alone. A possibility for testing a drug with profound effects on neurodegeneration would be a head-to-head comparison of the new drug with an approved therapy, e.g., IFNβ-1b or mitoxantrone, in patients with SPMS. In particular, IFNβ-1b has a very modest effect on disease progression in SPMS, which makes it possible, even with modest sample sizes, to demonstrate the superiority of a new drug.

As trials of neurodegeneration in patients with the progressive forms of MS using clinical endpoints are cumbersome, long-lasting, and expensive, there is an urgent need for MRI surrogates for neurodegeneration, as shown in Table 1. In short-term trials of RRMS, a potential endpoint could be the proportion of gadolinium-positive lesions that evolve into T1 black holes. Such trials could be performed as placebo-controlled trials with parallel groups of 50-100 patients and a trial duration of 6 months [30]. MTR is another sensitive MRI technique that could be used in trials of comparatively short-term duration, although with MTR it would probably be impossible

Table 1. Capability of MR measures to show disease-associated processes in MS

MR measure	Inflammation	Neurodegeneration	Neurorepair
T2 lesions	++	+	+
Gadolinium-enhancing lesions	+++	–	–
T1 black holes	+	++	++
Atrophy	–	+++	+
MTR	+	++	++
DT MRI	++	++	+
^{1}H-MRS	+	+++	++
Functional MRI	+	++	+++

Degree of association: – none; + weak; ++ moderate; +++ strong.
DT MRI, diffusion tensor MRI; ^{1}H-MRS, proton magnetic resonance spectroscopy; MTR, magnetization transfer ratio.

to distinguish between demyelination and axonal loss. MTR can be measured either as histograms in normal-appearing white matter, or the evolution of MTR in plaques can be studied [40]. A promising MRI technique is 1H-MRS, which can be used to measure changes in neuronal function; either changes within the whole brain, in normal-appearing white matter, in cortex, or in MS lesions. However, this technique needs to be evaluated in larger patient populations.

In progressive MS, changes in atrophy measures either in the brain and/or at the cervical spinal cord level would be most appropriate. Trials using atrophy measures as primary MRI endpoint should have a duration of at least 1 year, and probably 2 years [61]. If the new drug possesses anti-inflammatory properties in addition to the effect on neurodegeneration, the baseline measure should be postponed to month 6 in order to avoid falsely apparent decrease in brain volume during the first 6 months of therapy, due to the cessation of inflammation-derived edema. However, it seems that trials in progressive MS need to have clinical endpoints to support atrophy measures [58]. In fact, the ESIMS trial of intravenous immunoglobulin (IVIg) in SPMS patients showed a modest effect on brain atrophy, but had no effect on any clinical outcome measure [24].

Suggestions for Trials of Neurodegeneration

A short-term, phase-IIa trial of neurodegeration in patients with RRMS could be performed as a placebo-controlled add-on trial in patients treated with IFNβ-1a intramuscularly. Suggested inclusion criteria are patients who have been treated with IFNβ-1a and have experienced one relapse during the last 12 months on therapy. Including only patients with nine or more T2 lesions and at least one T1 black hole could enrich the patient population. The trial would have a parallel group design, with one group of patients treated with the new drug and IFNβ-1a intramuscularly and the other group treated with placebo and IFNβ-1a intramuscularly. Patients would be followed with monthly MRI for 9 months. The suggested primary endpoint would be new gadolinium-positive lesions appearing on MRI scans at 1, 2, and 3 months that evolve into T1 black holes during the following 6-month follow-up with monthly MRI. Secondary MRI endpoints could be the evolution of MTR of new lesions. In a short-term trial of 9 months, secondary clinical endpoints that might show a positive trend could be time to first relapse and the annualized relapse rate.

In a phase-IIb trial of longer duration, other MRI endpoints could be brain atrophy, T1 black hole volume, gray matter atrophy, and MTR in normal-appearing white matter. Disease progression at 24 months measured on EDSS and Multiple Sclerosis Functional Composite (MSFC) could be chosen as clinical endpoints that might show a positive trend and thereby lend clinical support to an effect on neurodegeneration.

Initiation of a phase-III trial would have to await the positive results of a phase-II trial. The design of a phase-III trial could be placebo-controlled (which, however, is not feasible in many countries) or, more likely, a combination trial in which the tested drug or placebo would be added to an approved drug that targets inflammation. Disease progression measured using EDSS would be the primary endpoint, and secondary clinical endpoints would include changes in cognitive functions and annualized relapse rate. The first secondary MRI endpoint should be brain atrophy, and other secondary endpoints could include the MRI endpoints mentioned above for phase-II trials.

Conclusions

Disease progression, either in terms of EDSS worsening or cognitive function decline, is the only appropriate clinical endpoint in studies of neurodegeneration. Such studies require very large patient populations followed over long periods of time and, hence, MRI measures that allow detection of neurodegeneration in a shorter time are highly warranted for future studies. In short-term trials, the evolution of gadolinium-positive lesions into T1 black holes could be a possible endpoint, and MT MRI, which has proven valuable to measure medium- and long-term MS-related changes, might be able to provide valuable outcome measures in clinical trials of patients with MS. Whole-brain 1H-MRS would be a rational MRI endpoint to measure changes in neuronal integrity and appears promising, but needs to be evaluated in larger patient populations. In long-term studies of neurodegeneration, atrophy measures seem to be the most appropriate MRI endpoints. Regarding future studies of drugs with effects on neurodegeneration, short-time phase-II "proof of principle" trials might use MRI as the primary outcome measure (either the development of T1 black holes or MTR), whereas phase-III trials need to have clinical disease progression as the primary endpoint supported by MRI endpoints, e.g., atrophy measures or MTR.

References

1. Comi G, Filippi M, Wolinsky JS (2001) European/Canadian multicenter, double-blind, randomized, placebo-controlled study of the effects of glatiramer acetate on magnetic resonance imaging-measured disease activity and burden in patients with relapsing multiple sclerosis. European/Canadian Glatiramer Acetate Study Group. Ann Neurol 49:290-297
2. Hartung HP, Gonsette R, Konig N et al (2002) Mitoxantrone in progressive multiple sclerosis: a placebo-controlled, double-blind, randomised, multicentre trial. Lancet 360:2018-2025
3. Jacobs LD, Cookfair DL, Rudick RA et al (1996) Intramuscular interferon beta-1a for disease progression in relapsing multiple sclerosis. The Multiple Sclerosis Collaborative Research Group (MSCRG). Ann Neurol 39:285-294

4. Johnson KP, Brooks BR, Cohen JA et al (1995) Copolymer 1 reduces relapse rate and improves disability in relapsing-remitting multiple sclerosis: results of a phase III multicenter, double-blind placebo-controlled trial. The Copolymer 1 Multiple Sclerosis Study Group. Neurology 45:1268-1276
5. PRISMS (Prevention of Relapses and Disability by Interferon beta-1a Subcutaneously in Multiple Sclerosis) Study Group (1998) Randomised double-blind placebo-controlled study of interferon beta-1a in relapsing/remitting multiple sclerosis. Lancet 352:1498-1504
6. IFNβ Multiple Sclerosis Study Group (1993) Interferon beta-1b is effective in relapsing-remitting multiple sclerosis. I. Clinical results of a multicenter, randomized, double-blind, placebo-controlled trial. Neurology 43:655-661
7. IFNβ Multiple Sclerosis Study Group and The University of British Columbia MS/MRI Analysis Group (1995) Interferon beta-1b in the treatment of multiple sclerosis: final outcome of the randomized controlled trial. Neurology 45:1277-1285
8. Trapp BD, Peterson J, Ransohoff RM et al (1998) Axonal transection in the lesions of multiple sclerosis. N Engl J Med 338:278-285
9. Charcot JM (1868) Histologie de la sclerose en plaque. Gazette Hopital (Paris) 41, 554-566
10. Prineas JW, Kwon EE, Cho ES et al (2001) Immunopathology of secondary-progressive multiple sclerosis. Ann Neurol 50:646-657
11. Bruck W, Stadelmann C (2003) Inflammation and degeneration in multiple sclerosis. Neurol Sci 24(Suppl 5):S265-S267
12. Bruck W (2005) The pathology of multiple sclerosis is the result of focal inflammatory demyelination with axonal damage. J Neurol 252(Suppl 5):V3-V9
13. Lassmann H (1998) Neuropathology in multiple sclerosis: new concepts. Mult Scler 4:93-98
14. Kutzelnigg A, Lucchinetti CF, Stadelmann C et al (2005) Cortical demyelination and diffuse white matter injury in multiple sclerosis. Brain 128:2705-2712
15. Prineas JW (1979) Multiple sclerosis: presence of lymphatic capillaries and lymphoid tissue in the brain and spinal cord. Science 203:1123-1125
16. Serafini B, Rosicarelli B, Magliozzi R et al (2004) Detection of ectopic B-cell follicles with germinal centers in the meninges of patients with secondary progressive multiple sclerosis. Brain Pathol 14:164-174
17. Kappos L, Moeri D, Radue EW et al (1999) Predictive value of gadolinium-enhanced magnetic resonance imaging for relapse rate and changes in disability or impairment in multiple sclerosis: a meta-analysis. Gadolinium MRI Meta-analysis Group. Lancet 353:964-969
18. Miller DH, Barkhof F, Frank JA et al (2002) Measurement of atrophy in multiple sclerosis: pathological basis, methodological aspects and clinical relevance. Brain 125:1676-1695
19. Barkhof F, van Waesberghe JH, Filippi M et al (2001) T(1) hypointense lesions in secondary progressive multiple sclerosis: effect of interferon beta-1b treatment. Brain 124:1396-1402
20. van Walderveen MA, Kamphorst W, Scheltens P et al (1998) Histopathologic correlate of hypointense lesions on T1-weighted spin-echo MRI in multiple sclerosis. Neurology 50:1282-1288
21. van Walderveen MA, Barkhof F, Pouwels PJ et al (1999) Neuronal damage in T1-hypointense multiple sclerosis lesions demonstrated in vivo using proton magnetic resonance spectroscopy. Ann Neurol 46:79-87
22. Leary SM, Miller DH, Stevenson VL et al (2003) Interferon beta-1a in primary progressive MS: an exploratory, randomized, controlled trial. Neurology 60:44-51
23. European Study Group on Interferon beta-1b in Secondary Progressive MS (1998) Placebo-controlled multicentre randomised trial of interferon beta-1b in treatment of secondary progressive multiple sclerosis. Lancet 352:1491-1497

24. Hommes OR, Sorensen PS, Fazekas F et al (2004) Intravenous immunoglobulin in secondary progressive multiple sclerosis: randomised placebo-controlled trial. Lancet 364:1149-1156
25. Panitch H, Miller A, Paty D, Weinshenker B (2004) Interferon beta-1b in secondary progressive MS: results from a 3-year controlled study. Neurology 63:1788-1795
26. SPECTRIMS (2001) Randomized controlled trial of interferon-beta-1a in secondary progressive MS: clinical results. Neurology 56:1496-1504
27. Rovaris M, Filippi M (1999) Magnetic resonance techniques to monitor disease evolution and treatment trial outcomes in multiple sclerosis. Curr Opin Neurol 12:337-344
28. van Walderveen MA, Barkhof F, Hommes OR et al (1995) Correlating MRI and clinical disease activity in multiple sclerosis: relevance of hypointense lesions on short-TR/short-TE (T1-weighted) spin-echo images. Neurology 45:1684-1690
29. Filippi M, Rovaris M, Rocca MA et al (2001) Glatiramer acetate reduces the proportion of new MS lesions evolving into "black holes". Neurology 57:731-733
30. Dalton CM, Miszkiel KA, Barker GJ et al (2004) Effect of natalizumab on conversion of gadolinium enhancing lesions to T1 hypointense lesions in relapsing multiple sclerosis. J Neurol 251:407-413
31. Rudick RA, Fisher E, Lee JC et al (1999) Use of the brain parenchymal fraction to measure whole brain atrophy in relapsing-remitting MS. Multiple Sclerosis Collaborative Research Group. Neurology 53:1698-1704
32. Rao AB, Richert N, Howard T et al (2002) Methylprednisolone effect on brain volume and enhancing lesions in MS before and during IFNbeta-1b. Neurology 59:688-694
33. Zivadinov R, Bakshi R (2004) Role of MRI in multiple sclerosis. I. Inflammation and lesions. Front Biosci 9:665-683
34. Filippi M, Rocca MA, Comi G (2003) The use of quantitative magnetic-resonance-based techniques to monitor the evolution of multiple sclerosis. Lancet Neurol 2:337-346
35. Agosta F, Rovaris M, Pagani E et al (2006) Magnetization transfer MRI metrics predict the accumulation of disability 8 years later in patients with multiple sclerosis. Brain 129:2620-2627
36. Cercignani M, Bozzali M, Iannucci G et al (2001) Magnetisation transfer ratio and mean diffusivity of normal appearing white and grey matter from patients with multiple sclerosis. J Neurol Neurosurg Psychiat 70:311-317
37. Rovaris M, Agosta F, Sormani MP et al (2003) Conventional and magnetization transfer MRI predictors of clinical multiple sclerosis evolution: a medium-term follow-up study. Brain 126:2323-2332
38. Fazekas F, Sorensen PS, Filippi M et al (2005) MRI results from the European Study on Intravenous Immunoglobulin in Secondary Progressive Multiple Sclerosis (ESIMS). Mult Scler 11:433-440
39. Richert ND, Ostuni JL, Bash CN et al (2001) Interferon beta-1b and intravenous methylprednisolone promote lesion recovery in multiple sclerosis. Mult Scler 7:49-58
40. Inglese M, van Waesberghe JH, Rovaris M et al (2003) The effect of interferon beta-1b on quantities derived from MT MRI in secondary progressive MS. Neurology 60:853-860
41. Rovaris M, Bozzali M, Iannucci G et al (2002) Assessment of normal-appearing white and gray matter in patients with primary progressive multiple sclerosis: a diffusion-tensor magnetic resonance imaging study. Arch Neurol 59:1406-1412
42. Schmierer K, Altmann DR, Kassim N et al (2004) Progressive change in primary progressive multiple sclerosis normal-appearing white matter: a serial diffusion magnetic resonance imaging study. Mult Scler 10:182-187
43. Davie CA, Hawkins CP, Barker GJ et al (1994) Serial proton magnetic resonance spectroscopy in acute multiple sclerosis lesions. Brain 117:49-58

44. Davie CA, Barker GJ, Thompson AJ et al (1997) 1H magnetic resonance spectroscopy of chronic cerebral white matter lesions and normal appearing white matter in multiple sclerosis. J Neurol Neurosurg Psychiat 63:736-742
45. Narayana PA, Doyle TJ, Lai D, Wolinsky JS (1998) Serial proton magnetic resonance spectroscopic imaging, contrast-enhanced magnetic resonance imaging, and quantitative lesion volumetry in multiple sclerosis. Ann Neurol 43:56-71
46. Brex PA, Gomez-Anson B, Parker GJ et al (1999) Proton MR spectroscopy in clinically isolated syndromes suggestive of multiple sclerosis. J Neurol Sci 166:16-22
47. Cucurella MG, Rovira A, Rio J et al (2000) Proton magnetic resonance spectroscopy in primary and secondary progressive multiple sclerosis. NMR Biomed 13:57-63
48. De Stefano N, Narayanan S, Francis GS et al (2001) Evidence of axonal damage in the early stages of multiple sclerosis and its relevance to disability. Arch Neurol 58:65-70
49. Mathiesen HK, Jonsson A, Tscherning T et al (2006) Correlation of global N-acetyl aspartate with cognitive impairment in multiple sclerosis. Arch Neurol 63:533-536
50. Narayanan S, De Stefano N, Francis GS et al (2001) Axonal metabolic recovery in multiple sclerosis patients treated with interferon β-1b. J Neurol 248:979-986
51. Sarchielli P, Presciutti O, Tarducci R et al (1998) 1H-MRS in patients with multiple sclerosis undergoing treatment with interferon beta-1a: results of a preliminary study. J Neurol Neurosurg Psychiat 64:204-212
52. Parry A, Corkill R, Blamire AM et al (2003) Beta-interferon treatment does not always slow the progression of axonal injury in multiple sclerosis. J Neurol 250:171-178
53. Rocca MA, Falini A, Colombo B et al (2002) Adaptive functional changes in the cerebral cortex of patients with nondisabling multiple sclerosis correlate with the extent of brain structural damage. Ann Neurol 51:330-339
54. Lee M, Reddy H, Johansen-Berg H et al (2000) The motor cortex shows adaptive functional changes to brain injury from multiple sclerosis. Ann Neurol 47:606-613
55. Reddy H, Narayanan S, Woolrich M et al (2002) Functional brain reorganization for hand movement in patients with multiple sclerosis: defining distinct effects of injury and disability. Brain 125:2646-2657
56. Blinkenberg M, Rune K, Jensen CV et al (2000) Cortical cerebral metabolism correlates with MRI lesion load and cognitive dysfunction in MS. Neurology 54:558-564
57. Blinkenberg M, Jensen CV, Holm S et al (1999) A longitudinal study of cerebral glucose metabolism, MRI, and disability in patients with MS. Neurology 53:149-153
58. Comi G, Filippi M (2005) Clinical trials in multiple sclerosis: methodological issues. Curr Opin Neurol 18:245-252
59. Goodin DS (2004) Disease-modifying therapy in MS: a critical review of the literature. Part I. Analysis of clinical trial errors. J Neurol 251(Suppl 5):V3-V11
60. Roxburgh RH, Seaman SR, Masterman T et al (2005) Multiple Sclerosis Severity Score: using disability and disease duration to rate disease severity. Neurology 64:1144-1151
61. Molyneux PD, Kappos L, Polman C et al (2000) The effect of interferon beta-1b treatment on MRI measures of cerebral atrophy in secondary progressive multiple sclerosis. European Study Group on Interferon beta-1b in Secondary Progressive Multiple Sclerosis. Brain 123:2256-2263

Subject Index

Printed in July 2007